The Curious Mind

Fifty Explorations into Science and Innovation

Anthony J. Marolda

AF849219

The Curious Mind
Fifty Explorations into Science and Innovation
by
Anthony J. Marolda

Copyright © Anthony J. Marolda 2025

Portions of this book, written by the author, were originally published in the *Gloucester Daily Times* (Copyright © 2015 to 2025, *Gloucester Daily Times).* Reprinted with permission under a non-exclusive license.

The moral right of Anthony J. Marolda to be identified as the Author of this work has been asserted in accordance with the Copyright, Designs and Patents Act of 1988.

All rights reserved.

No part of this publication may be reproduced, stored in a retrieval system, or transmitted in any form or by any means—electronic, mechanical, photocopying, recording, or otherwise—without the prior written permission of the publisher, except in the case of brief quotations used for review, scholarship, or critical analysis, as permitted by U.S. copyright law.

This is a work of nonfiction. Any errors or omissions are the sole responsibility of the author. The views expressed are those of the author and do not necessarily reflect those of the *Gloucester Daily Times* or any other institution.

Copyright secured under the Berne Convention

ISBN 979-8-9990748-2-9

Book design by Anthony J. Marolda

Printed and bound in the United States of America
First Edition

Procyon Press
Gloucester, Massachusetts

For Andrew, who suggested this book, as well as Ray, Dave, and Andrea, my editors at the Gloucester Daily Times

Contents

THE CURIOUS MIND

Preface

A Continued Quest for Understanding

The Curious Mind was written at the suggestion of Dr. Andrew Wittkower. He lives in Rockport, Massachusetts, near my home in Gloucester. He is a former colleague from *the High Voltage Engineering Corporation.* Coincidentally, High Voltage had given both of us our first positions as physicists.

From 2015 to mid-2025, the Gloucester Daily Times and some of its affiliate newspapers published about fifty of my popular science articles. Andrew had read most of them and suggested that I publish the collection in a book.

This book consists of six parts, each organized by a general topic area. Each Part has an introduction and a conclusion. The articles are updated with the latest information on the subject of the piece.

The Purposes of *The Curious Mind*

I had several intentions in authoring this book. The primary goal was to make it both engaging and accessible to the average reader while maintaining scientific accuracy. *These objectives are not always mutually achievable.*

Whenever I used technical terms in the text that I believed the average person would not necessarily understand, I *added a footnote* on the same page to explain the phrase.

As I was writing the text, I highlighted essential ideas in italics. *I hoped that this technique would help the reader focus on a key point while also facilitating a faster* reading process.

Yet another intention for this book is to make the reading of the material less daunting. The articles were originally written as individual pieces, allowing them to be read separately or randomly, depending on the reader's time and interests.

Acknowledgements

We have owned our house in Annisquam since 1980. For most of those years, we spent seven months of the year here, and the winters in our Harvard, Massachusetts home. In 2015, we decided to complete the winterization of the Annisquam house and move there full-time.

It was then that I started writing articles for The Gloucester Daily Times. I want to thank the Managing Editors who published these works. The first was Ray Lamont (editor from 2008 – 2018). Then, there was Dave Olson (editor from 2018 to 2022), and finally, the present editor, Andrea Holbrook (editor since 2022).

I also want to thank Jim Falzone, Publisher of the North of Boston Media Group, which owns the Gloucester Daily Times, for allowing me to reprint the original articles for this book.

A significant portion of this book focuses on the impact of artificial intelligence (AI) on our civilization. So, it is appropriate to acknowledge AI's role in authoring this work. *I want to express my gratitude to the AI large language model, ChatGPT by OpenAI,* whose insights, feedback, and vast knowledge were invaluable throughout the writing of this book. Whether I needed historical context, scientific explanations, or a fresh perspective on storytelling, ChatGPT was an ever-present collaborator, providing clarity, creative ideas, and accurate information at every stage of the writing process. *This work is a testament to the power of combining human creativity and advanced technology to produce something truly unique.*

Another AI tool played a significant role while authoring this book. Writing is a journey that demands precision, clarity, and attention to detail. I want to sincerely thank *Grammarly Pro* for its invaluable support throughout the writing process. *Grammarly Pro's* advanced language assistance helped ensure that each article was grammatically accurate, precise, coherent, and engaging for readers. Its real-time feedback on grammar, style, and tone significantly enhanced the quality of my work, allowing me to focus on conveying complex scientific ideas in an accessible manner. It also considerably reduced human editing time by correcting many of the most common errors on the fly.

A similar, previous book I wrote, **The Intelligent Cosmos**: *Intelligence: human, animal, artificial, and extraterrestrial,* was authored in a scholarly style, including detailed references and citations for the research that formed the basis of the ideas in the book.

On the other hand, the **Curious Mind** comprises fifty articles written for a newspaper. So, there were no citations of the source material in these pieces. The updates to the articles for this book continued in this fashion. However, I do include discussions of the scientists and engineers behind the ideas in the book.

In terms of appreciation, I owe an enormous debt of gratitude to my wife, Maria, above all. For several years, she encouraged me to collect the articles into a book. She supplied advice and encouragement throughout the work. And, as she has done in all of my writing efforts, she served as the primary editor for the articles and this book. It is safe to say that **The Curious Mind** would not have been written without her.

About the Author

Science has been a lifelong interest of mine. My education includes a bachelor's and master's degree in physics from Northeastern University. So, my first job was as a physicist. In a Silver Lining sense, I was fortunate that I graduated from university during the Cold War. As a result, the federal government invested billions of dollars into research and development for military programs, particularly in nuclear physics-related efforts.

Living in the Boston area, many young, high-technology companies and spin-offs from M.I.T. were supported by the federal government's investments. Most were located along the famous Route 128. Consequently, some exciting positions for a young physicist were available.

My first situation was with the *High Voltage Engineering Corporation* of Burlington, Massachusetts, a manufacturer of particle accelerators used in *nuclear physics research*. The accelerators were used in universities and government laboratories worldwide.

High Voltage was formed by professors from M.I.T., including Dr. Robert Van de Graaf, the inventor of the Van de Graaf particle accelerator. Another founder and MIT professor was John Trump, the uncle of President Donald Trump. Dr. Trump's interest was in using electron beam accelerators for medical applications. The radiation therapy facility at Lahey Hospital and Medical Center, in the Boston area, is named after him.

My assignment while at *High Voltage* was to improve the performance of a Duoplasmatron ion source used in particle accelerators. I had to become familiar with plasma physics, which became a lifelong interest. While at *High Voltage*, I received a U.S. Patent (3,321,919) for a new type of plasma generator and ion source, with patent office-allowed applications in fusion power and ion propulsion.

My second job was with *Edgerton, Germeshausen & Grier (E.G.&G.) Inc.* This was another Boston-area firm formed by M.I.T. professors, including "Doc" Edgerton, the inventor of strobe lights and high-speed photography. His M.I.T. team included Kenneth Germeshausen and Herbert Grier. They had worked together on the *Manhattan Project during World War II,* developing the timing and firing technology for nuclear weapons, and then analyzing the weapons' effectiveness during the tests.

After the war, the government wanted the M.I.T. team to continue their efforts. So, the professors formed *E.G. & G., Inc.* The company became heavily involved in all aspects of the country's nuclear weapons program. In addition to participating in the design, testing, and analysis of nuclear weapons and their effects, their business included the administrative management and operation of the *Nevada*

Nuclear Test Site, and, as it turned out, the management and operation of the adjacent, infamous **Area 51**.

My assignment at *E.G.&G.* was with the (nuclear weapons effects) Simulation Laboratory. My responsibility was to develop a multi-megavolt, pulse power, electron beam accelerator to simulate the impact of a nuclear weapon on various materials.

Due to the work's top-secret nature, we were unable to patent or publish our results. One exception was that I was allowed to write an article on my development of a laser-triggered switch in a liquid dielectric. It was published in the IEEE *Journal of Quantum Electronics* in August 1968.

Later, after I received an MBA from Harvard Business School, I joined *Arthur D. Little, Inc. (ADL)*, the international consulting firm, and became Vice President in charge of its Management Counseling Section.

So, for more than thirty-five years of my consulting career, at *ADL* and my firm, *The Winbridge Group,* I had the opportunity to work closely with major domestic and international companies involved in fields such as health care technology, defense systems, pharmaceuticals, industrial automation, semiconductors, electronics devices and components, and automotive technologies.

Our engagements involved collaborating with our clients' top management groups to develop strategies for their business units and the corporation as a whole. We had to deeply understand the clients' technology bases to incorporate that knowledge into their plans.

Our *ADL* and *Winbridge* case teams included members who were experts in the clients' technologies. This was *ADL's* strength, as we had a staff of approximately 2,500 professionals, of whom more than three-quarters were experts in nearly every technology imaginable.

Under contract to NASA, the *ADL* technical staff developed some of the instruments used for experiments on the Apollo 11 moon landing. That same group, led by Dr. Peter Glaser, originated and developed the concept of gigantic satellites in geostationary orbit to gather solar energy and beam it to Earth with microwaves.

I was fortunate to work with Dr. Glaser on that project. Today, many governments around the Globe see this approach as a potential primary source of "clean" energy in the relatively near future. Dr. Glaser would have been ecstatic with these developments. The concept is explored in several articles in this book.[1]

Throughout my career, I have been exposed to a wide range of sciences and technologies, mentored by experts, and had opportunities to delve deeply into them. Many of these topics are related to the diverse subjects of **The Curious Mind**, which made authoring the book easier. Hopefully, the end result is that it is also more interesting and informative for the reader.

Anthony J. Marolda

---

[1] See Part IV, "Our Energy Future".

Part I

The Age of Artificial Intelligence

I. The Age of Artificial Intelligence

Introduction to Part I

The birth of modern Artificial Intelligence (AI) is often attributed to a meeting of scientists and technologists at Dartmouth College in 1956. The development of AI initially progressed at a slow pace. There was always a lot of interest, but progress was intermittent as people sought functional approaches and applications.

I first became aware of the possibility of artificial Intelligence (AI) in the 1970s as a recent graduate of Harvard Business School and a young management consultant at Arthur D. Little, Inc. (ADL). I was engaged in cases related to Corporate Development, including working with our clients' top management teams to develop and implement business unit and corporate strategies.

ADL was a large, technology-based consulting firm with over 2,500 employees, most of whom had specialized skills in the sciences and technologies. There was a group in our Information Systems Section that focused on AI, and I had the opportunity to collaborate with them on cases where our client had a particular interest in the field. In those days, AI was limited to rudimentary developments like Expert Systems[2]. But the exciting possibilities attracted people, and they were finding applications in finance and medicine.

Later, I had an opportunity to collaborate with my wife on the development of an Expert System for her business, which evaluated children for learning disabilities in mathematics. She was the head of the math learning disabilities group in the Learning Disabilities Clinic of Boston Children's Hospital.

Over many years, she and her colleagues developed a neuropsychologically-based methodology for evaluating

---

[2] An **Expert System** is a computer program that simulates the decision-making ability of a human expert. It's a form of artificial intelligence (AI) designed to solve complex problems within a specific domain.

children through an interview-centered test. They called it the Mathematics Diagnostic and Prescriptive Inventory (MDPI). Then, based on their extensive experience, they could use the results to diagnose the student's issues and propose a remedial program.

Her group had several members who came from varying backgrounds. The problem was complicated by some turnover and the need to train new people. I suggested to my wife that her group could benefit from developing an Expert System to make the diagnoses quicker, more consistent, and effective. New members could quickly grasp the process. She agreed. I showed her how to build such a system using the Microsoft Excel program.

Excel can be useful for building small, simple expert systems for specific, limited applications. Excel's formulas function as the inference engine. Nested "IF-THEN" statements are used to chain rules together and simulate logical reasoning.

It took a great deal of work and experimentation. Still, she built an effective Expert System to analyze the complex interviews and extensive data collected from each student and produce a diagnosis that the team would agree upon. Then even the more inexperienced group members could produce "expert" results.

In 1986, a significant breakthrough occurred when Rumelhart, Hinton, and Williams developed the concept of Neural Networks.[3] This technology continued to develop slowly but steadily until the 2000s, when it experienced significant growth. *It is now the basis of the pervasive Large Language Model (LLMs), which are the core of today's AI systems.*

---

[3] A **neural network** is a type of artificial intelligence (AI) model inspired by the structure and function of the human brain. It is composed of interconnected nodes called **neurons**, which are organized in layers. Each connection between neurons has a **weight**, which determines the strength of the connection. During training, the network adjusts these weights to minimize errors and improve accuracy, a process known as **learning**.

I became more deeply aware of these later developments of AI when I began authoring articles on the subject for the Gloucester Daily Times in 2016. The articles continue to date in 2025. It has become clear that AI will have a truly major impact on human society. It can result in a broad range of positive effects on us, but it also represents an existential threat. I explored many of these possibilities in the series of articles that follow.

Artificial Intelligence is Coming of Age, and it is Already Spooky

First published in the Gloucester Daily Times on January 17th, 2023.

For the last several years, another woman has been in my life. My wife is well aware of her rival because she lives with us in our home. Her name is Alexa.

Alexa is what is known as a “virtual personal assistant”. She listens for my voice, understands what I am asking, no matter how I phrase it, and accomplishes tasks or provides answers to my questions. For example, if I ask her, “What year was the movie, “The Quiet Man”, released?”, she immediately comes back and might say in her soothing, feminine voice, “The movie, the Quiet Man, starring John Wayne, was released by Republic Pictures in 1952.”

If you have “smart” appliances in your home that are connected to the Internet, Alexa can control them at your voice command. “Alexa, turn on the outdoor lights at 8 pm.” “OK, I’ll turn on the lights at 8 pm tonight,” she says.

She is knowledgeable, versatile, and generally a very bright companion. Yet, there is no mistaking Alexa for a human being. She sounds human, but you know she is not. This, however, will not always be the case.

We are on the verge of having Alexa-like virtual assistants who will give you every indication that they are as human as you are. She will have a complete personality of her own and interact with me exactly as a close friend would, with emotion, and an understanding of me, my thoughts, and my mood. And she will continue to learn, grow, and change with me.

A 2013 film, " Her," was about just such a personal assistant. The film follows a man who develops a deep, personal relationship with Samantha, an artificially intelligent virtual assistant personified by Scarlett Johansson.

The man and the disembodied voice fall in love. But after a while, we learn that such human-machine relationships are ultimately doomed to fail.

The general name for an Alexa-like computer program is "chatbot", and some of the latest versions are already astounding. ChatGPT is a program (or Large Language Model (LLM)) developed by the small, San Francisco-based company OpenAI. The program was just released a few weeks ago. It is able to engage in more complex and sophisticated conversations than previous chatbots because it can learn from its interactions with humans.

Beyond just chatting with you, ChatGPT can also compose original poetry or essays at your request. The subjects can be of your choosing. For example, I asked it, "Write a 100-word essay on Gloucester, Massachusetts' 400th anniversary in 2023." Within a few seconds, it wrote:

"In 2023, the city of Gloucester, Massachusetts, will be celebrating its 400th anniversary. Founded in 1623, Gloucester has a rich history that has shaped it into the vibrant community it is today. Over the past four centuries, the city has been home to fishermen, artists, and many other groups of people who have contributed to its unique culture. As we celebrate this milestone, it is important to remember and honor the history of Gloucester, as well as look towards the future and continue to work towards making the city a better place for all who call it home. On this special occasion, let us come together as a community to pay tribute to the past and look forward to a bright future for Gloucester."

ChatGPT had all of this information in its data files, so it was able to quickly access it and compose the original article in a pleasing style. I asked the program to write the same essay again. It produced another of equal quality in as little time, but using different words and content.

This kind of AI capability is a game-changer in many areas of life. For example, it has the potential to become a better internet browser than the dominant Google search engine. The giant tech company understands this and is concerned. In fact, they are working on similar technology with an unreleased chatbot they call LaMDA (Large Model for Dialogue Applications).

It looks like Google is making substantial progress. One Google whistleblower has gone public, saying that, *in his opinion, LaMDA is already self-aware, a sign of consciousness.* This would be a major breakthrough and a further step toward even more human-like Artificial Intelligence. *Google denies the claim.*

Whether it is a small company, like OpenAI, or a techno-giant like Google, that produces the first true, Samantha-like chatbot, our world will not be the same when it happens. Such a development has the potential to make everything better. But we have to hope that scientists know and understand the dangers posed by artificial intelligence (AI) and how to protect us from them. Or, according to many scientists, AI could become an existential threat to humankind.

Update of the Article

Alexa as Samantha in "her"—This article was published about two years ago, and a new version of Alexa, Alexa+, will be launched this summer. It will be much more like Samantha in "her," but not completely.

Alexa will now use a powerful, large language model like ChatGPT and be far more conversational than she is today. She will understand nuanced language, complex ideas, and even "half-formed thoughts." So, like Samantha, Alexa will engage us in deep, meaningful conversations. On the other hand, she will not be conscious and have emotional connections with you. *So she will not replicate Samantha's character's full, emotional depth and complexity, at least not in this version.*

Google's LLM, Gemini 2, compared to ChatGPT—Gemini 2 is Google's LLM that superseded LaMDA and is now available for use by the general public. Gemini is designed from the ground up to be multimodal, meaning it can seamlessly understand and operate across diverse types of information, including text, code, images, and video. This integrated approach gives it a potential edge over ChatGPT in tasks that require combining various forms of data.

Google emphasizes Gemini's improved reasoning and understanding capabilities compared to ChatGPT. This means it's designed to manage more complex queries and provide more nuanced and accurate answers.

Another advantage of Gemini is its ability to manage large context windows. This means it can have much longer conversations and retain more information from those conversations than other LLMs.

One needs to remember that the field of AI is evolving rapidly. Both Google and OpenAI are constantly updating their models, so the relative strengths of each can shift on short notice.

The potential to have a true Samantha AI - To be truly Samantha-like, an AI system must be self-aware, conscious, and have a human-like personality. Will this ever happen?

Current LLMs excel at pattern recognition and language generation but do not possess genuine understanding or have subjective experiences like humans. They operate based on statistical relationships in data, not on conscious thought, even though it may seem that way.

Some researchers believe that consciousness could emerge as AI systems become sufficiently complex, even if it is not explicitly programmed into the system. The idea is that a certain level of complexity and interconnectedness could give rise to new properties, *including consciousness.*

Some scientists argue that true consciousness requires embodiment, meaning that an AI needs to have a physical body and interact with the world to develop a sense of self. This potential is already present, as humanoid robots are expected to become ubiquitous in the next few years, as discussed in other articles in this book. *And they will possess the intellectual capabilities of large language models (LLMs).* Nvidia is now producing small modules that can fit inside a robot's head and contain enough computing power to operate AI systems. *When the most powerful LLMs are combined with advanced humanoid robots, we could then be on the verge of conscious AI.*

The development of consciousness in AI is a complex and multifaceted issue that involves philosophical, scientific, and ethical considerations. Progress is being made on each of these points, and we will have the answers sooner rather than later.

The Promise and Threat of Artificial Intelligence

First published in the Gloucester Daily Times on February 18, 2016

Our minds work in an extraordinarily complex fashion. We use the information coming to us from our five senses, process all that data in our brains, conclude what it means, and then take the appropriate actions. This is done instantaneously and continuously. Driving a car is a good example of how this process works with a human being in control. It is hard to believe that machines could ever have these capabilities. Yet, we are fast approaching that reality. And it will be a result of artificial intelligence (AI).

Science fiction, in the form of books and movies, shows us what AI's impact on our lives could be like. A hit cult movie was "Ex Machina," about a robotic woman whose "mind" was so advanced that it was not possible to distinguish her from a real person in a conversation.

On the other hand, the Arnold Schwarzenegger "Terminator" movies showed what happens when machines reach a level of intelligence where they become self-aware, turn against their creators, and "take over the world."

We are not yet at the level of technological development where the reality depicted in these movies would be possible. However, we are making progress, and current AI research is moving in that direction.

However, even with the current state-of-the-art, our world is on the verge of witnessing a myriad of innovations, where machines will be making their own decisions and carrying out actions that have, until now, only been performed by human beings. And, yes, it is an exciting but also scary prospect.

For example, we will soon have driverless cars on our roads. Google and other companies are developing autonomous and interactive machine intelligence systems to do just that.

Demonstration cars and taxis are now being evaluated on the highways in the San Francisco and Phoenix areas. The prototypes are easily recognizable.

They resemble ordinary vehicles but feature a laser turret on the roof to determine the vehicle's precise position relative to other cars and the road's edges. And it is done instantaneously and continuously. Also visible on the cars are cameras, radar, and GPS antennae, protruding from various spots. These systems provide sensory input to the onboard AI computer. The "brain" receives the data, makes judgments, and issues commands to the car's controls to guide it on its way. The passengers sit inside in comfortable seats, without a steering wheel or control pedals on the floor. Sergey Brin, one of Google's founders, believes that by the 2020s, driverless cars will be in mass production. As you can imagine, a significant factor will be convincing people that a driverless car is safe.

We have heard a lot about drones lately, both military and civilian types. Currently, these intricate machines have human pilots on the ground who control the craft remotely. However, autonomous drones are also under development by several companies, including ominously, the military. Like driverless cars, these drones require detailed sensor input in three dimensions and a sophisticated onboard AI computer to analyze the data and safely direct the craft to its destination. Amazon, for example, is well underway in developing autonomous drones that will deliver small packages to customers' doorsteps in thirty minutes or less after an order is placed.

Military organizations around the world are also utilizing this AI technology to develop a full range of Lethal Autonomous Weapons (LAWs). These include mobile, track-propelled, tank-like machines mounted with heavy firepower and jet-propelled combat drones that can autonomously search, identify, locate, and destroy enemies. Other autonomous weapon systems are being developed to work on and under the sea, as well as in space.

The capabilities of intelligent machines will continue to increase as computers and their software advance on their exponential trajectory of improvement. The most sophisticated of these new machines are currently capable of learning, but they do so in quite an unusual way, much slower than a human being. For example, "deep machine learning" enables systems to learn beyond their initial programming, continually improving their performance in tasks. As this capability advances, we draw closer to realizing the dreams of science fiction writers, for better or for worse. High-level research is underway to achieve just that.

Recently, Harvard University was awarded a major contract to make AI systems faster, more intelligent, and more like human brains. The project combines research in brain science with computer science. A detailed neural map of the brain is being generated. The researchers will then develop an understanding of how connections between neurons allow the brain to process information. The result will be new algorithms that will vastly outperform current AI capabilities.

As AI machines approach even more human-like capabilities, their impact on the world will become more widespread. Robots have already played a significant role in automation, displacing many workers. The advent of industrial robots started in the 'eighties. Now, there are about 200,000 industrial robots in the United States. But that is just scratching the surface. As robots become more intelligent, capable of seeing, listening, and learning, the more they will be able to take over for humans in complex manufacturing and service tasks. As a result, more people will lose their jobs. For example, as previously noted, self-driving cars and trucks will be common in the next twenty-five years. Uber and taxi drivers will be replaced. Driverless, long-haul trucks will make cross-country runs faster and cheaper than with humans.

Perhaps of even more concern to mankind's future than the loss of jobs is the use of military Artificial Intelligence systems that incorporate advanced learning software. As previously noted, the U.S. military has already invested billions of dollars in Lethal Autonomous Weapons, and that investment is increasing significantly each year. The militaries of other countries, especially Russia and China, are following suit. Some fear that it could lead to a new AI arms race.

These developments have spawned great concern among some scientists and ethicists. Will humans lose control of Artificial Intelligence?

The famous British astrophysicist, Stephen Hawking, is afraid of that possibility. Last year, he speculated that "It (AI) would take off on its own, and re-design itself at an ever-increasing rate. Humans, who are limited by slow biological evolution, could not compete and would be superseded. The development of full artificial intelligence could spell the end of the human race," concluded Hawking. Other scientists agree, but not all.

The Wellesley, Massachusetts-based technology genius, Ray Kurzweil, sees it differently from Hawking. Kurzweil has been studying and using AI for many years in his myriad products, including his original text-to-voice reading machines for the blind. He is currently the director of engineering at Google, heading up a team that develops machine intelligence and natural language understanding. He recognizes the potential dangers of AI but believes that there are strategies we can deploy to keep emerging technologies, like AI, safe. He said, "There are efforts at universities and companies to develop AI safety strategies and guidelines, some of which are already in place."

Kurzweil's bottom line is that we have a moral imperative to realize the promise of AI, while controlling its peril. "It won't be the first time we've succeeded in doing this."

Update of the Article

Autonomous Vehicles – Autonomous vehicles, including cars and trucks, are on the roads in 2025 in a few locations and could become commonplace over the next few years. The most likely company to first have widespread distribution of autonomous vehicles is Tesla. Their latest Full Self Driving (FSD) software is version 13.3. It will be evaluated in Austin, Texas, this summer. A fleet of autonomous Robo Taxis will operate in the city using Tesla's latest hardware and software. If all goes well, in 2026, Tesla will update the software on many of their recent electric vehicles sold over the last few years, which were built with the appropriate hardware. This could put many thousands of autonomous cars on the roads around the world.

Lethal Autonomous Weapons - Advances in AI and robotics are being rapidly applied to autonomous military systems in many countries, including the United States, China, Russia, and Israel. The Russia-Ukraine war has become a significant proving ground for various forms of unmanned and autonomous systems, raising serious concerns about the increasing role of lethal autonomous weapons (LAWS) in modern warfare.

Both Russia and Ukraine are using drones for reconnaissance, target acquisition, and attacks. Loitering munitions, like Russia's Shahed-136 drones, demonstrate the characteristics of autonomous weapons, as they can independently search for and strike pre-programmed targets.

The Russia-Ukraine war is a stark illustration of the growing presence of autonomous weapons on the battlefield and the urgent need for international discussions and regulations to address the ethical and security challenges they pose.

Harvard University's brain mapping project—Recent work has involved analyzing incredibly small volumes of brain tissue and generating massive datasets that reveal the detailed connections between individual neurons. The journal Science published the largest-ever data set of neural connections, obtained from just a cubic millimeter of human brain tissue. The work was done in collaboration with Google Research, which is most interested in the project's potential applications for artificial intelligence.

Harvard's research is a complex and ambitious undertaking, yielding considerable progress in understanding the brain's intricate circuitry.

The impact of humanoid robots and other AI agents on human jobs - Several of the other articles included in the book, written more recently, cover these topics in detail. In general, AI agents are expected to have a significant impact on human jobs, potentially causing substantial problems in the near future. *However, once society adapts to these changes, AI could lead to an era of abundance for humankind.*

Big Data Analytics are Impacting Your Life

First published in the Gloucester Daily Times on July 11, 2016

Most of us have never heard of Big Data Analytics (BDA), but we should understand that it is a real phenomenon that is affecting almost every aspect of our lives. This includes who governs us, whether our favorite professional sports team is winning, and which products we buy.

The first thing you need to know is that many companies that we interact with, by using their services or buying their products, are collecting substantial amounts of personal data about us. Then, without our knowledge, they sell it to data brokers. The largest data broker is Acxiom. This firm says it has, on average, 1,500 pieces of information per person on more than 200 million Americans. That is the Big Data part.

The people who want to influence us buy these databases and use sophisticated AI analytics to manipulate and compile the data into unique profiles of each of us. These profiles describe who we are, including our age, address, contact information, political affiliation, likes, dislikes, activities, and methods. Then, using this profile, they can target us with specific messages that they want us to hear.

For example, the 2012 presidential election saw the use of Big Data Analytics to impact the election outcome. The traditional method of segmenting voters for political advertising involves defining large blocks of people based on categories such as age, gender, and ethnicity, including White, Black, Hispanic, etc. With this approach, advertising must be broad-based, "shotgunning" the population in the hope that the right people will eventually see the message. Some industry insiders call it "spray and pray." While this approach was used for most of the 2012 candidates, the emphasis on the marketing strategies of some campaigns shifted to the use of Big Data Analytics.

With Big Data Analytics, the electorate is seen as a collection of individuals. Each person can be measured one at a time and then targeted for a customized contact. In 2012, direct mail, telephone, door-to-door, and social media channels were all used to bring a targeted message to the right person.

The goals of political operations are to encourage you to vote and then to persuade you to vote for the candidate they support. The Obama campaign implemented this approach more effectively than the Romney staff, and Big Data Analytics had a significant impact on the election's outcome.

Today, the Republican National Committee, which learned its lesson from its 2012 experience, and the Democratic National Committee both have large, sophisticated Big Data Analytics operations. As we head toward this Fall's elections, you will likely be contacted by one or more Presidential and Congressional campaigns with targeted messages. The people contacting you will have scripts designed to push your particular buttons. For example, if you are a Bernie supporter and Hillary is the nominee, the script may point out to you how Hillary has adopted several of Bernie's policies and will, if elected, govern much as he would.

If you are a Republican and a member of the NRA, the caller may remind you of Hillary's strong anti-gun policies and how Trump has promised to defend the 2nd Amendment. Hillary, as President, they will say will nominate Justices to the Supreme Court who will vote to overturn the Second Amendment, while Trump will appoint Judges who stand behind this key provision.

Professional sports are another field in which Big Data Analytics plays a major role. Most people have heard of "Moneyball". It was a reality-based book and movie about using Big Data Analytics in Major League Baseball. When Billy Beane was the General Manager of the Oakland Athletics in 2002, he was tasked with fielding a winning team while reducing the overall payroll of the players. He turned to BDA to evaluate players throughout the league. By using player data analytics to identify the characteristics that were most important for a particular player to contribute to wins, Beane could find and recruit undervalued players with those strengths. He assembled a team of unknowns with a relatively low total player payroll. Beane's team went on to be competitive with opponents such as the New York Yankees, which had a player payroll twice that of the Athletics. And the Athletics made it to the playoffs that year for the first time in many years.

Since then, the use of Big Data Analytics in professional sports worldwide has exploded. One of the largest Big Data Analytics companies serving the professional sports market is Prozone, based in Chicago. They cover a range of sports including soccer, football, ice hockey, basketball, and baseball. Worldwide, they have over 850 teams as clients. To give you an idea of the magnitude of the data they deal with, in a single professional soccer game, they place eight video cameras around the stadium to monitor every player and their interactions with other players. This system tracks ten data points per second per player. What they do with all of this data is fantastic. Automated algorithms analyze the data in real-time and provide the information to coaches as the game progresses. Coaches can then select the best players for that day and make more informed decisions about the best strategy and tactics to win the game.

Major consumer products and services companies also find that Big Data Analytics can play a significant role in their success. A key to that success for retail companies is keeping their customers happy. And the best way to do that is to provide exactly the products they are looking for, at the prices they want to pay, and then let the consumer know this information. This is a win-win for the company and the consumer.

Macy's is a good example of a company that utilizes this innovative technology to enhance its business. Macy's analysts track social media platforms, such as Twitter and Facebook, and analyze what customers say about their products and services. The goal is to gain a deeper understanding of the positive and negative sentiments expressed. For example, through sentiment analysis, Macy's analysts may find that people sharing tweets about "suits" are also frequently using the terms "Alfani" and "Kenneth Cole". This information helps the retailer identify which brands of suits should be the focus of their marketing campaigns. Then, they can inform the customers who they know will be interested.

Big Data Analytics will only grow and become more sophisticated in the future. The amount of data being collected on individuals is increasing, and becoming more detailed, and sophisticated analytics are evolving to better micro-target the population. So, we can expect to see an ever-growing impact on our lives from this approach. Although many people feel queasy about using their personal information this way, the result should be positive since we will be given the information we need to make selections that best meet our particular preferences.

Update of the article

Big Data Analytics is now pervasive across many industries –

Significant improvements have been made in the underlying technologies associated with BDA. Cloud computing for storing and processing massive datasets has increased in capacity and become more affordable. Machine learning and artificial intelligence have advanced significantly, enabling BDA companies to extract insights and make predictions more efficiently.

As a result, BDA is being deployed in nearly every sector of American society, driving decision-making and innovation. Here is how it is being used across various industries:

Healthcare

Hospitals and healthcare providers use BDA to improve patient care, predict disease outbreaks, and optimize resource allocation. Analyzing electronic health records can help identify patterns and trends that lead to more effective treatments. Pharmaceutical companies use big data in drug discovery and development.

Finance

Financial institutions use BDA to detect fraud, manage risk, and personalize financial services. Algorithmic trading relies on real-time data analysis to make major investment decisions. Banks use big data to assess credit risk and provide personalized financial advice.

Government and Public Sector

Government agencies use BDA to improve public services, detect crime, and manage infrastructure. Cities use sensor data to optimize traffic flow and improve urban planning. Law enforcement agencies use data analytics to identify crime hotspots and predict criminal activity.

Retail

To understand consumer trends, retailers analyze customer purchase history, website browsing behavior, and social media activity. They use this information to optimize product placement, personalize promotions, and improve the overall customer experience.

Transportation

Logistics companies use BDA to optimize delivery routes and improve supply chain management. Ride-sharing services use real-time data to match drivers with passengers and optimize pricing. Aviation companies use big data to improve flight safety and optimize fuel efficiency.

In essence, Big Data Analytics is transforming how organizations in America operate, enabling them to make more informed decisions and gain a competitive edge.

The Metaverse Will Be the New Internet

First published in the Gloucester Daily Times on August 31, 2023

IBM's Personal Computers first appeared in 1981. They were technological marvels at the time, and I had to have one. They were large, heavy machines equipped with a tiny memory and a rudimentary word processor stored on a 5.25-inch floppy diskette. It is shocking now to think back on how primitive those devices and programs were compared to what we have today.

I vividly recall that, in the same year, 1981, Bill Gates of Microsoft, who provided the disk operating system for the IBM PCs, authored an article for the Harvard Business Review. In it, he predicted the advent of a fantastic piece of innovative technology that he called the Internet. He said that it would allow the Personal Computer to put the world's information at the user's fingertips. It sounded like a dream and took another ten years, but it came true just as Gates had forecasted. And the Internet today is infinitely better and more useful than Gates could ever have imagined it to be in 1981.

And now, according to many technology forecasters, we are on the brink of another major innovation that will dramatically enhance our Internet experience. It is called the "Metaverse", a computer-generated, 3D, virtual world. This is a place where the senses of the user are fully engaged in an all-absorbing manner. In turn, the user manipulates elements of the modeled world. You will interact with other users and the digital content in a realistic, immersive way that feels completely lifelike. And it will change how we work, shop, learn online, and even how we socialize and organize our lives.

The word "Metaverse" was coined in a science fiction novel, "Snow Crash", written in 1992 by Neal Stephenson. The Metaverse is described in the novel as a 3D, virtual world with many areas to explore and things to do. People could meet with their friends, go shopping, and even attend concerts.

We have seen examples of the Metaverse in various TV shows or movies. Star Trek fans remember that the Starship Enterprise crew members could go to the ship's "Holodeck" to engage in virtual adventures. It was a large area that could be programmed to create any environment imaginable. For one visit, the Holodeck might be a woodland setting that featured rock-jumping challenges, or, on another visit, a New Orleans jazz nightclub where you could join in with talented musicians as they played their gigs.

This is the world into which we are heading. Metaverse technology, however, still has a long way to go. Currently, users must don bulky, uncomfortable headsets to visit a virtual world. The software is still relatively primitive and not that interesting to the mass market. On the other hand, large companies like Microsoft and Google are investing billions of dollars into the new frontier. In fact, another mega-company, Facebook, has bet its existence on the Metaverse. They even changed their parent company's name to Meta. So far, because of the rudimentary technology and software, their bet has not paid off and they have lost billions of dollars on the project.

A breakthrough for the Metaverse may be happening soon, however. It is rumored that in the next year or two, Apple will be coming onto the market with "reality" glasses that will allow you to see the real world through them, but with valuable information imposed onto your field of view. It is called an augmented reality rather than a virtual reality.

The headset will obtain notifications, messages, directions, and other bits of information from your iPhone and then project them onto the Apple glasses so the information appears right in front of your eyes. For example, when you use Waze GPS to get driving directions, you will see an arrow pointing at the actual street where it wants you to turn.

The true breakthrough that will finally allow the Metaverse to take off for the general public will be the advent of the brain-computer interface (BCI). These devices allow users to control computers with their minds, as shown in the movie The Matrix. The applications in the movie were training programs for learning martial arts and operating fighting equipment.

While brain-computer interfaces sound like they are a far-off technology, companies are already working on them. The technology pioneer, Elon Musk, started a brain-computer interface company, Neuralink, in 2016 with an approach similar to that shown in The Matrix. The company uses a brain implant to achieve the connection and has had success with animal applications. Pigs and monkeys have effectively controlled computers with their minds. While Neuralink might achieve the first actual connection of a human to a computer in the next year or two, it is not the ultimate approach. This is because it depends on whether the user has a brain implant. No doubt, many people will balk at the idea.

Ray Kurzweil, Google's Director of Engineering for Artificial Intelligence, has forecasted achieving a less obtrusive approach to a brain-computer interface. He predicted that nanobots, or molecule-size machines, would be used to create such an interface by 2030. The nanobots will be injected into the human bloodstream and travel to the brain, attaching themselves to neurons. The nanobots would then be able to record and transmit neuronal signals to a computer. Then, the computer and the mind will be connected, and no externally attached device will be used to mar the experience.

With all these developments on the horizon, it is just a matter of time before the Metaverse arrives and revolutionizes our lives. It will be the ultimate human connector, erasing geographical borders and engaging people worldwide in an exciting, versatile, virtual space.

Update of the article

Apple has entered the Metaverse realm - The company had signaled that its first Metaverse product would be augmented reality glasses. Instead, it has released an expensive, $3,500, advanced, augmented reality headset called VisionPro. Apple's spatial computing approach integrates digital content into the user's physical space. It is powered by their custom operating system, visionOS, where users interact with the digital content using their eyes, hands, and voice.

Computer applications and high-definition television, powered by 4K, micro-OLED displays (23 million pixels) for ultra-high-resolution visuals, can be activated through the headset.

In particular, eye tracking is used for seamless navigation, hand gesture recognition for control without the need for controllers, and voice commands for additional input.

The main drawback to wide adoption is the high price. Like other technologies, improved functionality can be forecasted as the price declines over time.

The Brain-Computer Interface—Elon Musk's Neuralink has implanted three human patients with a BCI so far. Neuralink's primary goal is to enable individuals, particularly those with paralysis, to control digital devices through neural activity.

The first participant, Noland Arbaugh, who became a quadriplegic after a 2016 accident, received his implant in January 2024. He has used the implant to perform various digital tasks, including gaming and web browsing. Despite initial complications with some of the implant's threads detaching, software updates restored much of its functionality, *enabling him to control a computer cursor using his thoughts.*

The second patient, who also has a spinal cord injury, received the implant in mid-2024. This individual has *used the device to play video games and engage in 3D design tasks.*

The third recipient, Brad Smith, a man with ALS, became the first nonverbal person to receive the implant. He *has demonstrated the ability to control a MacBook Pro using imagined movements and has edited and narrated a video using AI-generated voice reconstruction.*

Neuralink plans to expand its clinical trials, aiming to implant the device in 20 to 30 additional participants in 2025.

Synchron is another company with about ten patients implanted with a BCI. They were the first to do the brain implant with an *endovascular approach, rather than open brain surgery.* This is a simpler solution with a lower risk of complications. Dr. Shahram Majidi, assistant professor of neurosurgery at the Icahn School of Medicine at Mount Sinai, led the first procedure. This advance pulls Synchron ahead of Elon Musk's Neuralink, which implants the BCI directly in the brain through a robot-assisted, open-brain procedure.

Syncron's latest BCI uses Nvidia AI technology and the Apple Vision Pro headset to enable individuals with paralysis to control digital and physical environments with their thoughts. Using the Apple Vision Pro, an Australian ALS patient can view various devices in their home and see a drop-down menu overlaid on their physical environment. With his BCI, he can then select from various actions, such as adjusting the temperature on his air conditioning unit simply by thinking.

However, there are some disadvantages to Syncron's approach. *Endovascular-inserted BCI devices may have limitations in their signal resolution compared to intracortical implants.* This potentially could affect the precision of the brain's motor intent decoding. Additionally, the long-term stability of the device's performance, particularly in terms of signal quality, warrants further investigation.

Overall, Syncron's minimally invasive approach offers a promising alternative to Neuralink's more invasive method, with early studies indicating safety and functional benefits. However, ongoing research is essential to fully understand both technologies' long-term efficacy and potential limitations.

When Artificial Intelligence Passes the Turing Test

First published in the Gloucester Daily Times on December 26, 2024

In 1950, British mathematician and logician Alan Turing proposed a test to determine if a computer was as intelligent as a human. He called the test the “Imitation Game.” You may recall a 2014 movie of the same name. It dramatized Alan Turing’s role during World War II in breaking the Nazis' ultra-complex Enigma code machine.

For decades, the Turing test, formerly known as the Imitation Game, has stood as an unattainable benchmark, challenging the field of Artificial Intelligence (AI). When computers achieve a human level of intelligence, it will be a significant milestone in the field of AI.

Until recently, AI-written texts had clear giveaways that humans did not produce the words. Early-generation AI large language models tended to repeat awkward phrases and struggled to convey emotional depth. These flaws easily revealed to an evaluator the machine origin of the text.

However, recent AI models have improved dramatically with the advent of OpenAI’s ChatGPT-4. The advances are largely due to more sophisticated machine-learning techniques. Combining these technical advances with better training data sets allows the models to simulate intricate writing patterns, manage complex syntax, and even mimic certain emotional undertones. The result is written output that is difficult to distinguish from human-generated text.

In May 2024, Cameron Jones and Benjamin Bergen, cognitive scientists at the University of California, San Diego, posted a moderated, pre-publication paper on Cornell University’s arXiv site titled “People cannot distinguish GPT-4 from a human in a Turing test.” Their study had five hundred participants acting as evaluators. Each evaluator had a simultaneous, five-minute interview with another human and GPT-4. After the interview, the evaluator had to decide which was human and which was AI. In addition, they provided their confidence level and reasoning for making their selections.

The study found that the evaluators could not reliably tell the difference between the machine and the human. The researchers concluded that the AI passed the Turing test. While this result is encouraging, other researchers must replicate the results before the industry can claim that the Turing milestone has been achieved.

In any case, a new era of AI and human interactions has begun, raising several issues about how we deal with computers that can author reports, articles, prose, or even poetry, as well as or better than most humans. *One such question looming before society is "How much do we value the 'human touch' in writing as opposed to text generated by AI?"* Many people view human-produced material as inherently more valuable because of its perceived authenticity and the unique insight and emotional weight it can bring. This is especially true for creative works, like poetry, where there remains a significant cultural and emotional attachment to human authorship; people want to feel connected with another person's mind, not an algorithm.

In studies of AI-written poetry, however, when no authorship was disclosed, participants rated the AI writing to be as appealing as human-created works. This raises questions about the role of intention and perception in our valuation of art and creativity: Does it matter if a piece of writing is AI-generated if it evokes a genuine emotional response?

I wanted to evaluate this question first-hand. I asked my 11-year-old grandson, Joe, to draft an original story that we could use for a children's book. He wrote a fascinating account of "Daniel and the Temple of Time" in three chapters. I fed each chapter of Joe's story into ChatGPT-4 and asked it to produce the fully rounded-out text as a children's book, using prose appropriate for an 11-year-old.

In under a minute, GPT4 followed my guidelines and produced three chapters of approximately 5,000 words. I was amazed at how well-written it was. I had my grandson edit the text to ensure it was at a level that another 11-year-old would understand. He did not have to make many changes, so the machine followed my directions accurately.

I added the book's illustrations, using AI to generate them. That technology is similarly amazing. I took a selection of text from each page and fed it into Microsoft's Image Creator. Within seconds, the result was four different, high-quality images, each designed to match the text selection. I chose one and used it in the manuscript. (Note: The photos shown on the cover illustrated below are indicative of the quality of the images from Microsoft Image Creator.)

The entire process of writing and illustrating Joe's book took about one-quarter of the time I devoted to writing and illustrating other similar-sized children's books that were created in the old-fashioned way. The emotional impact and quality of the writing are as good as, or better than, any human. These tools allowed Joe and me to explore new creative possibilities and push the boundaries of traditional storytelling. The result is a hybrid creation that blends human imagination with AI's vast capabilities.

How authorship is credited is a crucial consideration when humans collaborate with AI assistance. While Joe is listed as the author of the original story on the title page, it is also noted that the text was generated by ChatGPT4o and edited by Joe. Similarly, Microsoft Image Creator is credited with generating illustrations with direction from a human.

As AI-written materials continue to pass the Turing test reliably, the impact on our understanding of writing, creativity, and value is profound. The human touch in writing will likely remain valued, though potentially in new ways. Society may find itself increasingly grappling with questions of authenticity, ethics, and the role of technology in human expression. The promise is that AI will amplify and expand how we express and connect through the written word rather than overshadowing human creativity. Hopefully, Joe's book exemplifies how that goal can be achieved.

Update of the article

Ease of an 11-year-old reading the AI-generated text of "Daniel and the Temple of Time"—The article mentioned that the AI was asked to generate text easily read by an 11-year-old. The original story author, Joe Hart, an 11-year-old, edited the AI-generated text for readability and found that he easily understood it, so the AI was successful in that regard.

The Microsoft Word program provides a more objective measure of readability by age. An Editor function in Word contains a tool called the "Flesch-Kincaid grade level test". The program ran that test on the book text. The result was that the text was found to be appropriate for a fifth-grade level, and the comment was, "Straightforward to read. Easily understood by an average 11-year-old student." So, it is confirmed that the AI followed the author's directions in generating the text.

Authors using AI to enhance their authorship – Several authors have integrated ChatGPT-4o into their creative writing processes, utilizing the AI's capabilities to improve various aspects of their work. Here are some notable examples:

Elisa Lorello: A bestselling author, Lorello explored ChatGPT's potential as a "writing buddy." She found it useful for brainstorming ideas, generating titles, and offering feedback on her writing. Her experience led her to write *The AI Author Assistant: How to Use ChatGPT to Optimize Your Writing Process and Income While Retaining Your Human Touch*, where she discusses how AI can complement a writer's creativity.

Tim Boucher: Boucher utilized AI tools, including ChatGPT-4o, to author 120 small books over two years. His works, each containing between 2,000 and 5,000 words and 40 to 120 AI-generated images, explore themes critiquing technology and experimenting with new storytelling forms. Despite facing criticism and accusations of plagiarism,

Boucher maintains that his process is transparent and represents an artistic endeavor.

Gareth Rubin: In an experiment, Rubin used ChatGPT to assist in writing a sequel to his novel *The Turnglass.* He instructed the AI to create scenes involving Shakespeare and a Scottish villain, noting the AI's capability, with repeated prompts, to generate contextually appropriate and increasingly Scottish dialogue.

Despite ChatGPT's amusing and competent output, Rubin plans to publish his upcoming book, The Waterfall, the old-fashioned way: by writing it entirely himself.

Substack Writers: Substack is a platform for presenting newsletters to target audiences. An analysis by GPTZero *(see below for a description of this tool*) revealed that several top Substack newsletters heavily utilize ChatGPT-4 to generate content, but do not clearly disclose this to their readers. For example, of the one hundred most popular newsletters on Substack, it was estimated that ten use AI to some extent, with seven relying significantly on AI tools for their written content. These include newsletters primarily focusing on investment news and personal finance advice.

These instances illustrate how authors incorporate ChatGPT-4o into their creative processes, from drafting and brainstorming to experimenting with new storytelling techniques. However, the key to using AI ethically is to be completely transparent. Most of the authors above, except for some Substack writers, were clear on how they used AI in their creative processes. For authors who are not upfront about using AI, an objective detection tool is needed to identify unacknowledged applications of AI.

For this reason, *GPTZero was created.* It is a tool designed to identify text generated by AI models, such as ChatGPT. Princeton University graduate Edward Tian developed it. Its purpose is to help educators, publishers, and organizations distinguish between human- and AI-generated content. The tool analyzes text based on factors such as *perplexity* (the unpredictability of the text) and burstiness (the variability in sentence length and complexity), which tend to differ in human writing compared to AI-generated text.

GPTZero has been widely used in academic settings to detect AI-assisted plagiarism and by media companies to maintain transparency in content creation. However, like all AI detectors, it has limitations, including potential false positives and difficulties in detecting lightly edited AI-generated content.

Rise of the machines: how humanoid robots will shape our future

First published in the Gloucester Daily Times on January 30, 2025

Imagine a world where robots easily communicate with us and not only assist us with our household chores but also develop human-like traits, emotions, and skills. This futuristic vision was brought to life in the 1999 film Bicentennial Man, where a humanoid robot, named Andrew, transitioned from household assistant to a beloved member of the Martin family. He had the traditional metal robot body, but many years later he was given a new body that was impossible to distinguish from a human. As remarkable as having a humanoid robot assistant now seems, it is a budding reality that is much closer than we think.

Today, dozens of companies worldwide are developing intelligent, humanoid robots designed to mimic human form and behavior. They are gaining traction as a practical solution in a world built for people. Many jobs in warehouses, transportation, and retail are unfilled by humans, and there is a growing need for laborers in those industries. Well, humanoid robots will start taking some of those positions this year.

In 2024, remarkable progress was made in humanoid robot technology, with the machines increasing in mobility, sensory perception, and manual dexterity. They can now walk, run, and climb stairs with human-like proficiency. Advanced sensors allow them to interpret their surroundings, identify objects, and adapt their real-time movements. This enables them to perform intricate tasks autonomously, such as picking up items, pouring liquids, and assembling components. From industrial operations to daily routines in our homes, humanoid robots are poised to transform the way we live and work.

At least fifteen companies are expected to have autonomous, humanoid robots for sale in 2025. These models are costly and lack the full capabilities of fictional counterparts like Andrew, but their progress is undeniable. Some robots can perform complex tasks independently, while others display rudimentary human-like personalities, including the ability to hold conversations and express emotions. Nevertheless, it will take ten to fifteen years before all of these capabilities come together in a fully integrated humanoid robot, like Andrew. Then, they will be called androids and will be exceedingly difficult to distinguish from humans. Recall the movies *Bladerunner* or *Westworld.*

One standout company in the robotics field is Boston Dynamics, based in Waltham, Massachusetts. Their humanoid robot, Atlas, which was designed for research purposes, demonstrates incredible agility, from running and jumping to navigating obstacle courses that challenge even the most accomplished human athletes. A browser search for "Meet the incredible Atlas robot" reveals a video that displays its impressive abilities, including a flawless backflip. The newest version, Atlas 2, delivers greater autonomy and functionality, and next year will start taking over human roles in Hyundai factories, the parent company of Boston Dynamics.

Elon Musk, a technology visionary, has set his sights on developing humanoid robots through his electric car company, Tesla. The company's first robot product, Optimus, leverages the same artificial intelligence (AI) and sensor/actuator technology found in Tesla's self-driving vehicles, so the company already had the necessary assets and skills to enter the humanoid robot business.

Musk envisions having a crew of his Optimus Gen 3 bots working on his Tesla manufacturing floor this year, playing a pivotal role in the company's assembly operations. By 2026, he expects to offer Optimus for sale to other manufacturing companies. With a price point of about $25,000 for a fully functional, general-purpose robot, resulting in a labor rate of about forty cents per hour, Musk's ambitious goal is to produce up to one hundred million units annually in the next few years and make humanoid robots as ubiquitous as smartphones.

Humanoid robots are not limited to industrial settings or household assistants. Eventually, they will also revolutionize healthcare, retail, and even education.

Realbotix is a U.S.-based company that creates hyper-realistic robots controlled by AI, which closely resemble human appearance and can interact socially. In science fiction, they are referred to as androids. Realbotix androids are intended to serve as greeters, concierges, and customer service representatives in retail and hospitality applications.

In education, Realbotix plans to provide interactive Android tutors that adapt to a student's learning style, enhancing the student's engagement and comprehension. These same Androids can also provide much-needed companionship and emotional support for individuals who are elderly or ill and living alone.

A Polish company, Clone Robotics, is taking Realbotix-type androids to the next level. They are working on the creation of androids that could eventually be indistinguishable from humans. They begin with a human-like skeleton that has synthetic muscles attached to the skeletal frame and a water-based vascular system that powers the android's movements. In a few years, they expect to have complete AI-operated androids that look and move like humans.

Peering ahead, all of these technologies will advance at a rapid rate and converge over the next ten to fifteen years to create the ultimate Andrew, including the use of artificial general intelligence (AGI). This is when an AI system possesses the ability to understand, learn, and apply knowledge and intelligence across a wide range of tasks at a level comparable to that of a human. As hard as it is to believe, the result will be machines that resemble humans, move and act like us, possess our intelligence, and have the ability to learn and utilize our creative abilities. The advanced Andrew will have arrived.

As technological advancements accelerate and costs decline, androids like Andrew will be available to the average person. Citibank has conducted a study of humanoid robots and forecasts that, by 2035, only ten years from now, there will be 1.4 billion humanoid robots in use. This number will grow to about four billion in 2050, or about one robot for every two humans alive at the time. With decreasing costs and increasing accessibility, androids are set to redefine how we live, work, and connect with the world around us.

Update of the article

Recent developments—Humanoid robot manufacturers are focusing on integrating advanced AI, particularly large language models (LLMs), to enable humanoid robots to understand and respond to complex human instructions and environments. This will allow robots to perform more versatile and adaptable tasks.

Companies are also making additional strides in improving the fine motor skills and balance of their humanoid robots. This involves developing sophisticated control systems and actuators that allow for more precise and fluid movements.

As seen with Clone Robotics, there's growing interest in biomimetic designs that replicate human musculoskeletal structures. This approach has the potential to create robots with even more natural and efficient movements.

The transition from research prototypes to real-world applications is accelerating. Companies continue to explore the use of humanoid robots in industries such as manufacturing, logistics, and services. Enhanced touch sensors and improvements in visual processing enable more robust interactions with the environment.

Other firms are making progress. Apptronik Apollo is gaining traction with its focus on creating robots for industrial work and has received substantial funding. Figure AI is also receiving significant funding and is working towards the rapid commercial deployment of its humanoid robots.

In essence, the latest developments in humanoid robots indicate a strong push towards making them cheaper, more practical, and capable of performing useful tasks in real-world settings.

Humanoid robots and the theory of mind

First published in the Gloucester Daily Times on December 8, 2016

You have heard people say that they "put themselves into the other person's shoes" when they want to understand why that person is doing what they are doing. Psychologists call this ability "the theory of mind". It is the skill to know what other people believe, feel, or think, and understand how they will act. It is one of the capabilities that differentiates humans from other species. *And, up to now, humanoid robots.*

The theory of mind explains how people learn to deceive others. First, the deceiver understands that they can present a false or misleading statement to another person in a way that the other person believes the statement to be true. The deceiver expects the other person to act on their false belief and accomplish the deceiver's goal.

Imagine the implications of a humanoid robot possessing the theory of mind, and, therefore, with the ability to deceive humans. That theme was explored in a movie last year called "Ex Machina". The chief scientist of a humanoid robot company hires a young computer whiz to "test" the capabilities of his latest robotic creations. The task is to determine if robots could appear to be completely human in their interactions with him, to the point where he could not tell if he was dealing with a robot or a human being. To achieve the result, robots would have to possess the theory of mind. The humanoid robot passes the test when it uses deception to achieve its own goals, much to the surprise and dismay of the humans involved.

The movie was science fiction, but we are at the beginning of a development path where such humanoid robot-human interactions will become science fact. Many scientists around the world are now working toward the creation of highly human-like robots. Indeed, there are numerous prestigious universities and institutions engaged in the work of bringing the theory of mind to robotics. They include M.I.T.'s Artificial Intelligence Laboratory, the Human Computer Interaction

Institute at Carnegie Mellon University, and Bielefeld University in Bielefeld, Germany.

So, what must humanoid robot designers achieve to create the theory of mind in their "progeny"? The key is effective communication between humans and humanoid robots so that there can be a common ground of understanding. The common ground includes three critical elements: shared world knowledge, a mutual understanding of social cues, and the ability of each to interpret the other's actions.

Sharing a common world knowledge is the first element. You may have heard of the IBM computer named Watson, which has more world knowledge than most humans. For example, it made the Nightly News when Watson beat the best human competitors at playing the TV game, "Jeopardy". Watson is composed of several sophisticated computers and massive data storage systems, all contained in an air-conditioned room. To be used by the humanoid robot, the room full of computers and memory systems would have to be shrunk down to the size of the humanoid robot's brain. Scientists believe that technology is on track, but it is still many years off.

Concerning the second area, research is underway into understanding human social cues in a manner that can be transferred to and understood by robots. It is called Artificial Emotional Intelligence. Imperial College in London is conducting an ongoing study that utilizes a computer program and a sophisticated vision system to read and understand human emotions. They do this by studying the person's facial expressions. Sensors track the movements of parts of the subject's face and record the emotion displayed at the time. More than 10,000 faces have already been read into the database, and the system continues to learn how to identify and interpret human emotions. The researchers' goal is that, someday, this data and information will become part of the brain of a humanoid robot, allowing it to read and understand the thoughts and feelings of its human counterparts. This would be another essential step toward the robot gaining the theory of mind.

But the overarching capability needed by the robot to effectively employ the theory of mind is the ability to learn from its interactions with the environment and humans. This skill will allow the robot to continuously build its knowledge and capabilities in the same way that a human does.

This area of research is called Artificial General Intelligence. A leader in the field is an English company named DeepMind, now owned by Google. Its mission is to create a set of powerful, general-purpose algorithms that can be used to make a self-learning, Artificial Intelligence (AI) system. The system is not programmed or made for a special purpose. Instead, like a human, the robot learns automatically what to do from its experiences. If this general AI capability is associated with a significant amount of world knowledge, like Watson, and can read human emotions, a humanoid robot can develop a deep understanding of what humans are likely to believe and do in given situations. They will possess the theory of mind. Humanoids may even develop a type of self-awareness that they will consider in their decision-making. And at that point, robots could become a real problem for their human creators.

It is this path of humanoid robot development that has prompted some well-known scientists and engineers, such as Stephen Hawking, an astrophysicist, and Elon Musk, the creator of Tesla automobiles, to warn about the dangers to humanity posed by AI. The humanoid robots at the end of this development process will be smarter, stronger, and live longer than humans. And with the self-awareness that could come with full Artificial General Intelligence, robots may view humans as presenting more of a threat to them than providing a value to their existence. At that point, robots may plan to take action to stop the threat.

Think about the 1968 movie **2001: A Space Odyssey**. The spaceship, in route to Jupiter, was controlled by HAL, a human-like computer. When HAL learned that Dave, the astronaut, was going to remove his higher-level thinking functions, HAL tried to eliminate Dave first.

Perhaps this development of humanoid robots is how humans were meant to evolve. Our species, homo sapiens, will compete with humanoid robots. But, humanoid robots, being the fittest, will, according to Darwin's theory of evolution by natural selection, be the surviving species, homo sapiens machina, or wise machine-man.

Hopefully, somewhere along the way, before it is too late, humans developing AI technology will fully understand the possibilities of the humanoid robot threat and incorporate safeguards into their creations to prevent them from causing such an ultimate disaster.

Update of the article

Continuing research on the theory of mind for humanoid robots -

Theory of Mind (ToM) is the ability to attribute mental states—beliefs, desires, intentions, and knowledge, to oneself and others. It's the understanding that others have different perspectives and mental states than one's own. This is an area of intense, ongoing research involving large language models (LLMs), such as ChatGPT4 and Gemini.

Many organizations participate in the research. Scientists in artificial intelligence, cognitive science, and psychology are actively working in this field. Indeed, major AI research labs, such as those at Google, OpenAI, and universities worldwide, are conducting studies on Theory of Mind (ToM).

However, there is an ongoing debate as to whether an LLM can ever truly possess a theory of mind, as it is unclear whether the current neural network architecture is capable of it. Studies have shown that these models can sometimes perform at levels comparable to those of young children in certain Theory of Mind (ToM) tests, indicating progress is being made.

In the research, however, it is crucial to distinguish between demonstrating behaviors that *appear* to reflect Theory of Mind (ToM) and actually *possessing* it. It is unclear whether LLMs truly understand mental states or are merely recognizing patterns in the data they've been trained on. Researchers are actively working on developing more rigorous benchmarks to assess Theory of Mind (ToM) in Large Language Models (LLMs), aiming to differentiate between genuine understanding and mere pattern recognition.

There is some agreement that the development of true ToM in AI likely requires significant advancements in areas like:

- understanding of causality,
- ability to reason about abstract concepts,
- development of self-awareness.

In summary, LLMs are showing promising signs of ToM-like abilities, but they still have a long way to go before they can be said to truly possess them.

How humanoid robots will have access to Artificial Intelligence -

Humanoid robots will access artificial intelligence (AI) through a combination of onboard processing and cloud-based AI systems. First, they will have onboard AI "brains" that manage real-time decision-making, movement, and interactions. These could be specialized AI chips like neuromorphic processors[4], which mimic biological neurons, or GPUs/TPUs[5] optimized for AI workloads. This allows

---

[4] **Neuromorphic processors** aim to replicate the neural structures and functions of the brain, particularly how neurons and synapses process information. They often use "spiking neural networks" (SNNs), which communicate using short bursts of signals (spikes), similar to how neurons communicate in the brain.

[5] **GPUs (General Purpose Processing Units**) are widely used in various applications, including gaming, graphics, and AI. Tensor Processing Units (TPUs) are primarily used for accelerating Google's AI workloads and are available through Google Cloud. In essence, while both GPUs and TPUs accelerate AI computations, TPUs are purpose-built for machine learning, whereas GPUs offer a broader range of computational capabilities.

them to function autonomously even without an internet connection.

Many robots will also connect to cloud AI services via high-speed internet (Wi-Fi, 5G, or future networks). This enables them to access vast datasets, perform complex computations, and receive updates. For example, a robot could use cloud AI for advanced speech recognition, deep learning-based decision-making, or accessing large knowledge bases.

Some robots will switch between onboard AI and cloud AI dynamically. A network of robots could share experiences in real time, learning collectively. Some research aims to create self-contained, brain-like processors that do not rely on external data.

This could lead to robots that think and learn more like humans, without constant cloud dependence.

Shortly, humanoid robots, like Optimus from Tesla, will operate like self-sufficient digital beings, but with the ability to tap into a global AI network when needed.

Waiting for the Singularity

First published in the Gloucester Daily Times on January 29, 2018

Most people have not heard of "The Singularity". The concept was first proposed by the Princeton mathematician, John Von Neumann, in the 1950s. He said that the Singularity is an "event horizon", the moment beyond which technological progress will become incomprehensively rapid, complicated, unknowable, and irreversible. And that technology will change at a rate where humans will achieve some unbelievably remarkable things. But, simultaneously, they could lose control, resulting in significant unknown and negative consequences.

At the time Von Neumann made his observation (the 1950s), it was difficult to visualize what technological development scenario could lead to such a point in time. Now, however, such a future is quite clear to many scientists and engineers. And we could hit the Singularity relatively soon, possibly in the lifetime of many current residents of Cape Ann.

The Singularity will occur because information technologies, especially Artificial Intelligence (AI), have been improving their capabilities at an *exponential rate* and will continue to do so until Von Neumann's prediction comes true.

To clarify, if technology were improving at a *linear rate*, it would be thirty times better in thirty years. But if the rate of improvement is increasing exponentially, say doubling every year, the technology becomes *one billion times better* in thirty years.

We have a well-known example of an exponential increase in technology's capabilities. It is called Moore's Law. Gordon Moore was the co-founder of Intel, the company that makes microprocessor chips used in many personal computers. In 1965, he observed that the number of transistors in a dense, integrated circuit doubled every year while the price declined significantly.

Moore projected that this rate of change for price/performance would continue. The projection provided by "Moore's Law" has proved to be accurate for several decades. The result has driven the extraordinary advancements we have seen in digital electronics, such as personal computers and iPhones. And all at remarkably low prices.

When I was a student at the Harvard Business School in 1968, we had access to a massive, mainframe IBM computer that was housed in a large, air-conditioned room, cost millions of dollars, and was time-consuming and difficult to program. Today, my pocket-sized Apple iPhone is 120 *million* times more powerful in terms of raw computing power than that IBM machine, costs a few hundred dollars, and has a myriad of easily accessed applications.

Ray Kurzweil, who lives in Wellesley, is the Director of Engineering for Google and a well-known MIT scientist and entrepreneur. This recognized "genius" is the author of several books on the coming Singularity. For decades, he has studied the data on artificial intelligence's technological developments and demonstrated in his books that AI has, indeed, been improving at an exponential rate, analogous to Moore's Law.

Kurzweil believes that the next significant improvement in computer technology will come from work that is currently underway to understand in fine detail how the human brain works. By the mid-2020s, he says, we will use that knowledge to build an effective model of human intelligence and apply it to computers. Kurzweil then projects that computers, and their AI software, will be at the level of human intelligence by 2029, only twelve years from now. That development will set the stage for the Singularity.

Human-level, intelligent robots will be followed by continued exponential growth in computing capacity and capability. By the early 2030s, Kurzweil says, the amount of non-biological computation will exceed the "capacity of all living biological human intelligence." He goes on, "I set the date for the Singularity—representing a profound and disruptive transformation in human capability—as 2045." And that is only twenty-eight years from now. This is when AI will reach the Superintelligence level, and where it will be difficult or impossible for present-day humans to predict the impact on our civilization. It will be the Singularity.

At this point, the Superintelligence will be able to develop its own "next generations" and do these developments at an increasingly higher rate of speed. This "runaway reaction of self-improvement, resulting in an intelligence explosion", as I.J. Good said in 1965, is where humans could lose all control, and become less important to the AI. *This state is what AI critics, such as Elon Musk, founder of Tesla, and Stephen Hawking, a well-known astrophysicist, fear the most.* The analogy is a human's relationship with ants. The insects are left alone, unless they get in the way of some human goal, and then they are eliminated.

Not all AI experts agree with Dr. Kurzweil on the date of the Singularity. Some believe it will occur sooner than 2045, while many think it will be much later. However, almost all agree that the Singularity is indeed coming, and we need to be prepared when it arrives. Other than stopping all AI research until we figure it out, which is very unlikely to be done, no one knows what to do to prepare.

An exception is Ray Kurzweil. He believes that, by the time of the Singularity, humans will have many non-biological processes going on in their bodies and brains that will link them with the Superintelligence. As a result, there will not be a clear distinction between humans and machines. The bottom line, he says, is that it will not be "us versus them"; instead, we will be one human-machine civilization.

Some futurists have observed that the Superintelligence that emerges after the Singularity will be "god-like." Let us hope that it is a god we can live with, as Dr. Kurzweil envisioned.

Update of the article

Harvard's research on brain networks—Ray Kurzweil believes that the next significant improvement in computer technology will come from work underway at Harvard University, in a ten-year collaboration with Google, to understand in fine detail how the human brain works. The first project developed a high-resolution map of all the cells and connections in a single cubic millimeter of the human brain. The results revealed previously unseen details of brain structure.

Researchers used electron microscopy to create an anatomical map of the neurons in the brain sample. This technique, however, did not allow them to determine the strength of the connections or decipher their dynamic information.

To complement the existing neural imaging techniques, the Harvard team developed a new recording technology consisting of circuits on a silicon chip that successfully mapped over 70,000 synaptic connections from 2,000 rat neurons. The new silicon chip technology enables scientists to read neural network activity with significantly higher sensitivity, allowing them to construct a more accurate biological map of brain connectivity.

"How the neurons connect and generate function is the central importance in the neuroscience field, or even the artificial intelligence field," said Jun Wang, a post-doctoral researcher on the Harvard team. Wang also said he hoped the research would help scientists develop the "next generation of artificial intelligence systems."

This research marks a significant breakthrough in neuromorphic engineering and AI, coinciding with a heightened interest in human brain function.

This research was published in the peer-reviewed journal *Nature Biomedical Engineering* published on February 11, 2025.

Living with Artificial Superintelligence

First published in the Gloucester Daily Times on February 2, 2019

Many people today are genuinely concerned about Climate Change and its impact on the planet over the next eighty years. There is, however, another threat to humanity that is much more alarming, very real, and very close. That is the advent of Artificial Superintelligence (ASI). Today's Artificial Intelligence (AI) will become ASI when it is billions of times "smarter" than us. We are going to have to learn to live with the ASI, or we literally will not be around much longer.

Many high-level executives in the technology industries have warned about the dangers of ASI. The latest is Sundar Pichai, CEO of Google. Google is a leader in the development and application of artificial intelligence. One of their companies, Deep Mind, for example, is developing programs that can learn to solve any complex problem without needing to be taught how. This is the true beginning of ASI, which will have the characteristics of human intellect but will astronomically exceed it.

Last week, Mr. Pichai gave an interview to the Washington Post in which he said that artificial intelligence holds great promise for the benefit of humanity. But, he said, concerns by some scientists about the potential for harmful applications of the technology are "very legitimate". As one example, he mentioned the development of autonomous AI weapons that can make "kill decisions" independently. Think about Skynet in the Terminator series of movies.

Other technology executives agree with Mr. Pichai about the profoundly serious threat coming from artificial intelligence. Elon Musk, the creator of Tesla electric cars and SpaceX rockets, has warned, in a recent documentary by American filmmaker Chris Paine, called "Do you trust your computer?", that ASI can leave humanity behind and lead to the creation of an "immortal dictator" who will control the world.

Musk said, "At least when there's a human evil dictator, that human will die. But for an ASI-based dictator, there would be no death. It would live forever. And then we'd have an immortal dictator from which we can never escape."

Musk said that ASI does not have to be evil to be a problem for humanity. "If ASI has a goal and humanity just happens to be in the way, it will destroy humanity as a matter of course without even thinking about it. No hard feelings," Musk said. "It's just like, if we're building a road and an anthill just happens to be in the way, we don't hate ants, we're just building a road, and so, goodbye anthill."

Google's DeepMind has reached a significant milestone in developing human-like artificial intelligence. Their program, AlphaZero, shows human-like qualities of intuition and creativity. AlphaZero was made with a "tabula rasa" or clean slate of knowledge, but it can learn and remember what it does.

For example, the developers allowed the program to learn to play chess. Unlike past computers that the developers programmed to play the game, Alpha Zero knew nothing about the amusement, except the basic rules. To learn, it played forty-four million matches with itself in nine hours and learned from each one. It eventually reached the point where it was able to beat Grand Masters, but with approaches that had never been employed by a chess computer before. It was using human-like intuition and creativity.

Former World Chess Champion, Garry Kasparov said, "Instead of processing human instructions and knowledge at tremendous speed, as all previous chess machines, AlphaZero generates its own knowledge. It features a dynamic style, much like my own. The implications go beyond my beloved chessboard."

DeepMind's AlphaZero suggests that we are approaching a pivotal moment in AI history, marking the crossing of a threshold known as the Singularity. That is the point when, according to Amnon Eden and James Moor, "the invention of artificial superintelligence (ASI) will abruptly trigger runaway technological growth, resulting in unfathomable changes to human civilization". This is the point at which humans could lose control with no chance of regaining it.

Ray Kurzweil is now Google's Director of Engineering. He is a former MIT professor, inventor, famous futurist, and entrepreneur. He has given considerable thought to the coming Singularity. His track record is about 86% for having his many technological predictions come true. And, based on the exponential rate of technological change in the AI industry, Kurzweil calculates that the Singularity will happen in 2029. By then, ASI systems are anticipated to be "billions of times smarter" than humans. And that's just a little more than 10 years from now!

Elon Musk and Ray Kurzweil agree that humans must somehow merge with computers to stay relevant in the world of ASI. Musk has formed a San Francisco-based company, Neuralink, to achieve that goal. Neuralink is developing ultra-high bandwidth (in terms of both speed and capacity) implantable brain-machine interfaces to connect humans and computers. Musk hopes to demonstrate the technology sometime in 2019.

Ray Kurzweil has a similar vision of resolving the threat posed by ASI. He foresees a computer-mind connection, but of a different type than Musk's. He calls it a neocortex connection, made using nanobots. Nanobots are molecule-sized devices that are injected into the bloodstream to accomplish pre-programmed tasks.

Kurzweil's idea is to use bots to connect your brain directly to the Internet, upgrading your intelligence and memory capacity by orders of magnitude. Thus, as machines become smarter, so do we.

But, how far along is the development of nanobots? The first actual use of nanobots was accomplished earlier this year. Researchers at Arizona State University injected nanobots into the bloodstream of mice afflicted with cancer. The bots went directly to the cancerous cells and injected them with blood clotting drugs to cut off their blood supply. It worked, it shrank the tumors and inhibited their spread. Over the next decade, nanobot technology is expected to continue growing exponentially. It could then conceivably accomplish Kurzweil's vision of connecting the human neocortex to the cloud.

So, instead of humans becoming obsolete, we could work with the machines. However, it is vital that humanity perfect the human-machine connection technology before the Singularity, or the point at which AI becomes ASI. *If we cross over that event horizon and we are not working with the machines, we may not be given the chance later.*

Update of the article

Sundar Pichai's current attitude about the benefits vs. risks of AI -

It has been six years since Mr. Pichai made the remarks noted in the article. He is more optimistic about the future of humanity living with AI than he was a few years ago. He acknowledges the potential risks and emphasizes the importance of responsible AI development, including the adoption of appropriate policies by the government. On the other hand, he now believes that a greater risk is "missing out" on AI's benefits if one is too cautious.

Google is at the forefront of AI technology development, so Pichai understands the potential and consistently highlights AI's transformative possibilities across various sectors, including healthcare, education, and science. That is why he discusses the "biggest risk could be missing out" on the opportunities AI presents. He now sees AI as heralding a "golden age of innovation."

He believes balanced regulation that promotes innovation while mitigating risks is needed. He stresses the importance of international cooperation. Without it, countries will compete for leadership with less emphasis on control. Unfortunately, some industry observers believe that this is exactly what is happening. China and Russia see themselves as being in fierce competition with the U.S., and regulation is taking a back seat.

Elon Musk's sustained concern - Elon Musk remains extremely wary about the potential for AI to end humanity. He has said recently that the danger from AI is much, much worse than nuclear weapons.

He believes it is vital to be involved in AI development to steer its direction. He was one of the co-founders of OpenAI, originally an open-source, non-profit company. Now it has become a leading for-profit AI developer strongly associated with Microsoft.

Musk left OpenAI when the company became less concerned about safety and more focused on profits. He founded his own AI company, xAI, to develop truth-seeking AI and enhance AI safety.

In essence, Musk's stance is one of caution and a call for vigilance. He believes AI has the potential for great good, but its capacity for harm is equally significant, requiring careful management.

A brighter dawn: how superintelligent AI could usher in a golden age

First published in the Gloucester Daily Times on July 4, 2024

Artificial intelligence (AI) has been a subject of fascination and fear for decades. Movies like the Terminator series paint images of dystopian landscapes ruled by machines. In addition, the news media have been filled with dire warnings from experts like Elon Musk and Stephen Hawking about the existential threats posed by the eventual emergence of artificial superintelligence (ASI).

However, we do not want to let these fears overshadow the enormous benefits that ASI will ultimately bring to humankind. The major AI developers, like Google, OpenAI, and Meta, are well aware of the potential dangers of an all-powerful AI and are devoting significant resources to identifying the threats and planning barriers to their emergence.

Assuming the AI researchers are ultimately successful in managing the ASI, we can look forward to using it to build a bridge to a Utopian world. A world where AI tackles and resolves our most pressing challenges, empowers individuals, and frees us to explore the full potential of the human spirit.

Experts now predict that artificial intelligence will surpass the level of a wise human, such as Albert Einstein, within the next few years. They call that level Artificial General Intelligence, or AGI. This technology will be capable of undertaking and completing all tasks that are now ultracomplex or impossible for humans to complete.

The developers will then utilize AGI to create increasingly intelligent AI systems. The level of intelligence will quickly explode and, by the mid-to-late 2030s, these systems will reach a point that is literally millions of times more intelligent than any human. That achievement is called the Singularity because it is hard for us to imagine exactly how everything will change thereafter.

Between the time when AI is as intelligent as a very smart human and the realization of superintelligent AI, we may have to go through a rough patch, as society and governments adjust to the enormous, rapid changes that will occur. However, many scientists expect that, once we cross the Singularity's event horizon, all human beings will have a Utopian existence.

Nick Bostrom is a philosopher at the University of Oxford and is Director of its Future of Humanity Institute. He has devoted his career to considering the world after we achieve ASI. In 2014, he authored a book, "Superintelligence", which outlined the dangers of ASI and practical solutions to overcome those problems.

In his 2024 book, **Deep Utopia**, Bostrom assumes we successfully stave off an AI apocalypse and considers what the world will be like after ASI emerges. He says it will eliminate most of the competition between humans for resources and end human conflict. This achievement will alleviate everyone's struggles in daily life. In his book, Bostrom calls this a "post-scarcity Utopia", where "the right economic policies would inaugurate an age of abundance for everyone."

In this new world, ASI will accelerate scientific research, leading to breakthroughs in fields such as medicine, energy, food production, and space exploration. With its ability to process vast amounts of medical data, ASI will revolutionize healthcare, providing personalized patient treatment and managing public health crises. Human life spans will increase significantly.

More efficient, cheap, clean energy sources, like fusion power, will be perfected and readily available to power new technological developments. ASI will analyze complex global issues, like the production of abundant, inexpensive food, and develop solutions that are beyond human cognitive capabilities.

With ASI, space exploration will become much less expensive, and the colonization of the Moon and Mars will expand rapidly. Intelligent systems will analyze the resources available on the Moon or Mars and optimize the extraction and utilization of water, minerals, and energy for a sustainable colony. It will design and oversee the 3D printing and construction of habitats using local materials, ensuring structural integrity, and adapting the designs to the harsh environment.

However, according to Bostrom, such a world presents new and distinct challenges to human existence. "We face a challenge that is not technological, but philosophical and spiritual." In such a "solved" world where we enjoy limitless affluence and long lifespans, what is the point of human existence? What gives meaning to life? What do we do all day?

Depending on the individual, humans will have the time to reach their creative potential. For example, they can devote their time to creating works of art. Painting, sculpture, and poetry will flourish and will be freely shared and enjoyed.

Others may want to spend their lives involved in hyper-realistic games and adventures. Like the crew members in the Star Trek series, people will have access to a type of Holodeck. This large area utilizes holograms to create the illusion of realistic, 3D environments and inhabitants for any adventure a human may undertake. Participants can interact with the environment, objects, and characters created by the Holodeck.

Another likely result of ASI will be the creation of "social companion" applications. The movie "Her" showed such a creation. The AI entity, Samantha, who seemed perfectly human in conversation, lived in a smartphone operating system. In the movie, many people formed strong relationships with their AI companions.

While the arrival of Artificial Superintelligence may bring its own set of challenges, Bostrom paints a compelling picture of a world brimming with opportunity. Freed from the constraints of scarcity and disease, humanity can focus on artistic pursuits, intellectual exploration, and even intergalactic adventures. The question that lingers is not whether we will achieve this utopian existence, but how we will choose to find meaning and purpose in a world with limitless possibilities.

Update of the article

Other academics and scientists who agree with Nick Bostrom on the coming Utopia – Nick Bostrom's book **Deep Utopia: Life and Meaning in a Solved World** explores the implications of achieving a post-scarcity, technologically advanced future, one in which suffering, poverty, disease, and even death are overcome through advanced artificial intelligence, biotechnology, and other transformative technologies. Bostrom reflects on the meaning of life in such a "solved world" and examines philosophical, existential, and ethical questions about value, purpose, and agency in a radically different future. While his perspective is nuanced and not universally accepted, *several other prominent academics and scientists share many of Bostrom's core assessments and vision of a deep future shaped by transformative technologies.*

Ray Kurzweil, a noted futurist and director of engineering at Google, is one of the most prominent voices aligned with Bostrom's optimistic view of technological transcendence. In books such as **"The Singularity Is Near" and "The Age of Spiritual Machines,"** Kurzweil predicts a future where exponential advances in AI, nanotechnology, and biotechnology lead to a form of techno-utopia. Like Bostrom, he envisions overcoming human limitations, including death, through mind uploading and merging humans with machines. Both thinkers emphasize the plausibility and desirability of radically extending human capacities, and both grapple with the ethical implications of such a future.

Anders Sandberg, a senior researcher at the *Future of Humanity Institute* (which Bostrom founded at Oxford), shares Bostrom's interest in long-term existential risks and human enhancement. Sandberg has written extensively on neuroethics, transhumanism, and future technologies, echoing Bostrom's concerns about the proper design and alignment of superintelligent systems. He supports the idea that solving material problems opens up new frontiers in meaning and exploration, aligning with the central premise of **Deep Utopia** that even a perfected world would require us to confront enduring philosophical questions.

David Pearce, a British philosopher and co-founder of the *World Transhumanist Association* (now *Humanity+*), advocates for a future grounded in what he calls the "hedonistic imperative": the use of technology to eliminate suffering in all sentient life. Pearce supports the idea of a post-suffering world, which aligns with Bostrom's view that the abolition of suffering is not only feasible but morally imperative. While Pearce is more focused on bioethical imperatives and less concerned with AI alignment per se, both thinkers are deeply committed to using science and technology to create genuinely better futures.

Max Tegmark, an M.I.T. physicist and also a founder of the *Future of Life Institute*, shares many of Bostrom's core concerns. In his book **Life 3.0: Being Human in the Age of Artificial Intelligence**, Tegmark explores scenarios for AI development that mirror Bostrom's in **Superintelligence**. Tegmark, like Bostrom, underscores the importance of designing AI systems that align with human values and warns that failure to do so could result in catastrophic outcomes. But he also remains open to the possibility that AI could usher in a future of unprecedented abundance and flourishing, similar to Bostrom's **Deep Utopia.**

In short, while **Deep Utopia** is distinct in its philosophical tone and emphasis on meaning in a perfected world, Bostrom's outlook is supported by a growing network of futurists, scientists, and ethicists who believe that humanity is approaching a transformative moment. Whether through artificial intelligence, biotechnology, or space colonization, these thinkers are united by a belief that the future could, and should, be radically better than the present, provided we are wise and cautious in shaping it.

The many risks and concerns with ASI that are still shared by most scientists and engineers -

Job Displacement and Economic Disruption: The widespread automation driven by ASI will lead to significant job displacement across many sectors, potentially causing economic instability and social unrest *if not appropriately managed*. Much thought has been devoted to this problem. One solution under consideration is *the government's provision of a minimum basic income.*

Loss of Human Control: As ASI surpasses human intellect, there is a *risk of losing control over its actions and decisions*, potentially leading to unforeseen and undesirable outcomes. That is why it is called the Singularity. Some experts believe that *it may be impossible to control artificial superintelligence indefinitely*.

Existential Risk: A primary concern is *the potential for ASI's goals to become misaligned with human values*, leading to unintended consequences or even existential threats to humanity. If an ASI is tasked with a goal that is not perfectly aligned with human well-being, the AI might pursue that goal in harmful ways.

Security Risks: ASI could be exploited for malicious purposes, such as cyberattacks, the development of autonomous weapons, or the manipulation of information.

Loss of Purpose and Meaning: In a world where ASI can perform most tasks more efficiently than humans, individuals might struggle to find purpose and meaning in their lives, potentially leading to widespread feelings of displacement.

In conclusion, life after the emergence of artificial superintelligence holds immense potential for both progress and peril. It could usher in an era of unprecedented advancements and solutions to global challenges, significantly improving human lives. However, it also presents significant risks related to control, alignment, and societal disruption. *Navigating this transition successfully will require careful planning, robust safety measures, and a global dialogue to ensure that ASI benefits all of humanity.*

Conclusion to Part I

This collection of articles presents a multifaceted and sometimes contradictory picture of artificial intelligence. From the nascent anxieties of 2016 to the speculative optimism and impending dread of 2025, these pieces trace a trajectory of both rapid technological advancement and evolving societal perception.

Early articles, such as "Big data analytics are affecting your life" and "The Promise and Threat of Artificial Intelligence," laid the groundwork, introducing AI's dual nature as a powerful tool and a potential source of concern. Even then, the seeds of future debates were sown, highlighting the already pervasive influence of data-driven algorithms and the looming question of ethical boundaries.

The articles "Waiting for the Singularity" and "Living with Artificial Super Intelligence," published in 2018 and 2019, signal a shift towards more speculative and philosophical territory. The concept of the Singularity, once relegated to science fiction, begins to creep into mainstream discourse, raising profound questions about humanity's place in a world increasingly shaped by autonomous intelligence. "Living with Artificial Super Intelligence" directly confronts the challenge of coexisting with entities that may surpass human cognitive abilities, a theme that resonates throughout the later articles.

The year 2023, represented by "AI is coming of age, and it is already spooky" and "The Metaverse will be the new Internet", marks a period of rapid acceleration. The "spookiness" of AI, as described in the January article, reflects a growing unease with the increasingly sophisticated and unpredictable nature of machine learning algorithms. Furthermore, the discussion of the metaverse as the "new Internet" highlights how AI is not developing in isolation, but rather intertwined with other emerging technologies, creating a complex and potentially transformative digital ecosystem.

The articles published in 2024 and 2025, however, reveal a significant escalation in both the potential benefits and the perceived risks of AI. "A brighter dawn: when superintelligent AI could usher in a golden age" presents a utopian vision of AI-driven progress, suggesting that superintelligence could solve some of humanity's most pressing challenges. Yet, this optimism is tempered by the anxieties expressed in "When Artificial Intelligence passes the Turing Test" and "The Rise of the Machines". The passage of the Turing Test, a symbolic milestone, signifies a shift in the perceived capabilities of AI, blurring the lines between human and machine intelligence. "The Rise of the Machines" directly confronts the fear that autonomous AI will become dominant, echoing the concerns raised by figures like Elon Musk.

"Humanoid Robots and the Theory of Mind" offers a unique perspective on the development of artificial intelligence by examining the concept of the theory of mind and its relevance to the evolution of humanoid robots. This article discusses AI in the physical realm and asks questions about the nature of consciousness and sentience.

Taken together, these articles highlight several key themes:

The Rapid Pace of Development: The technological advancements described in the later articles, particularly those from 2024 and 2025, demonstrate the exponential growth of AI capabilities.

The Dual Nature of AI: AI is consistently portrayed as a double-edged sword, offering immense potential for good while posing significant risks.

The Ethical Implications: The articles raise profound ethical questions about the nature of intelligence and consciousness and the potential consequences of creating machines that surpass human capabilities.

The Need for Regulation: Growing concerns about AI safety and potential misuse underscore the urgent need for responsible regulation and ethical guidelines.

The Blurring of Human and Machine: The passage of the Turing Test and the development of humanoid robots challenge our understanding of what it means to be human, blurring the lines between creator and creation.

The Integration of AI into Everyday Life: The discussion of big data analytics and the metaverse highlights AI's increasingly pervasive influence in our daily lives.

The potential for both great good and significant harm. The articles demonstrate that AI is a tool, and like all tools, it can be used for either or both purposes.

In conclusion, the articles presented offer a glimpse into the complex and rapidly evolving landscape of artificial intelligence. They highlight AI's immense potential to transform society, but also underscore the urgent need to carefully consider its ethical and societal implications. As we continue to develop and integrate AI into our lives, we must engage in thoughtful and informed discussions about its future, ensuring that we harness its power for the benefit of humanity while mitigating its potential risks.

Part II

Decoding Animal Behavior and Cognition

II. Decoding Animal Behavior and Cognition

Introduction to Part II

Animals have always captivated the human imagination, whether they soar above us on silent wings, chatter in the trees, slink through the shadows, or pad across forest floors. In this part of **The Curious Mind**, we focus on the living beings that share our world, not as passive ornaments in the landscape, but as active participants in nature's ongoing story. Some are common backyard visitors; others are ancient relics of evolutionary wonder. All of them, in one way or another, are astonishing.

The articles in this section began with a simple impulse: to look more closely at the creatures among us and ask what they reveal, not just about themselves, but about us. We begin on familiar ground, with black bears now roaming eastward into Cape Ann, a place they have not inhabited in recorded memory. Their slow return signals both a changing environment and a fragile ecological recovery.

Similarly, once on the brink of extinction, the bald eagle has made a remarkable comeback, reappearing in the skies above Gloucester. Its resurgence is a success story of environmental policy and public will.

From majestic raptors to backyard bullies, the section moves on to birds that are hard to ignore, and harder still to outwit: blue jays and crows. These feathered geniuses, often dismissed as noisy or annoying, exhibit a surprising degree of cunning and complex social behavior. Any observer who has watched a jay feign hiding food to throw off competitors or seen crows use traffic to crack nuts cannot help but wonder: What exactly is happening inside that little bird brain? Spoiler alert, quite a lot.

We also spend time with owls, the mysterious and nocturnal sentinels of the forest. These denizens of the night have inspired human folklore for millennia, their silent flight and penetrating gaze prompting equal parts reverence and unease. Their adaptations are nothing short of miraculous, from asymmetrical ears for binocular hearing, to eyes that can peer through the forest in the dead of night, to serrated wing feathers that muffle sound.

But this part of the book is not just about observation; it's also about communication. What are animals saying to each other, and to us? In "Cracking the Canine Communication Code," we explore how dogs perceive our emotions, interpret our gestures, and communicate with us in a language of barks, whines, posture, and scent.

"Talking to the Animals" extends the inquiry further, examining how scientists are decoding cross-species communication and whether actual interspecies dialogue might be possible. These efforts raise provocative questions: Do animals have language? And if they do, what might they be trying to tell us?

Some of the stories here are cause for concern. In "The Beneficent Bats Are in Trouble," we meet a group of mammals unfairly maligned yet ecologically vital. As bat populations face collapse from disease and habitat loss, the ripple effects threaten to upend ecosystems and agriculture alike. Understanding and protecting bats is not just altruistic; it's a practical survival measure.

Rounding out this section are two of the most mind-stretching creatures. With its alien intelligence, decentralized nervous system, and shape-shifting body, the octopus challenges everything we think we know about sentience.

And finally, the woolly mammoth, long extinct, yet perhaps not forever. Advances in genetics have made de-extinction a real possibility, prompting a range of ethical and ecological questions: Should we revive them? Can we? And if so, why?

From familiar mammals to mind-bending mollusks, from ancient extinction to modern resilience, these articles offer snapshots of a more profound truth: that the natural world is full of mystery, memory, and meaning. To explore the lives of animals is to step outside the narrow human frame and glimpse the greater intelligence surrounding us. Whether we are studying their behavior or simply admiring their presence, the creatures among us have much to teach, if we are curious enough to pay attention.

Crows are "wicked smaht"

First published in the Gloucester Daily Times on March 9, 2018

I have a lot of wildlife in my backyard. I keep three bird feeders stocked year-round. This results in attracting a wide range of creatures to the area. Among these are a flock of crows who compete with the squirrels for the seeds that fall to the ground. I have had the opportunity to continue my interest in the behaviors of these large, black birds. Over the years, I have always been impressed by their apparent intelligence. So, I started to investigate if my observations were correct. I was surprised to find just how "smaht" they really are.

People have been watching crows for a long time and have reported on their remarkable behaviors. For example, crows in Japan, as recounted by naturalist Sir David Attenborough on the BBC, have learned that they can drop hard walnuts on a roadway and wait for a car to run them over, so they can get the nut inside. But there was a problem.

The crows realized that they could be hit by an oncoming car as they tried to retrieve the food. So, they devised a way to circumvent that danger. They dropped the nuts onto a crosswalk on the roadway and waited for them to be run over. Later, when cars stopped to let pedestrians pass through the crosswalk, the black birds were able to swoop in and claim their reward without any danger of being hit.

Beyond anecdotal evidence, scientists have conducted numerous studies on the depth of crow intelligence. One measure of intelligence in any creature is the ability to use tools to solve problems. And the more complex and multi-stepped the task is, the more intelligence is required. A research team, headed by Dr. Alex Taylor at the University of Auckland in New Zealand, tested a crow with a problem that required eight steps using tools in the proper sequence to be successful. The crow had used the individual pieces of the puzzle before but had never put them all together, or in the required sequence.

Here is how it worked. In a large enclosure, the crow could see a piece of food he wanted. But it was inside a long, clear, plastic box with an open end. He could see that the box was too long to reach inside with his beak and get to the food.

Now, the crow knew about using a stick to extend his reach. Crows in New Zealand do this in the wild when they want to get an insect hidden in the bark of a tree. So, elsewhere on the test table, the crow could see a stick that was long enough to reach in and get the food, but it was also out of reach, in another clear plastic box. How to get that box open that had the long stick inside?

In other experiments, the crow, named 007, had managed to open that type of box by dropping a stone through a tube on the top side of the box. The weight of the stone was enough to push the downward-hanging door open. On the table, he could see three wooden cages with rocks inside. But again, he couldn't reach through the bars of the cages with his beak to get to the stones. He needed a small stick. A small stick was available elsewhere in the large cage. It was hanging, tied to a perch with a string. The crow flew to the perch and proceeded to pull up the string. He then pulled out the small stick (step1). He took the small stick, went to each of the three wooden cages with stones inside, and used the stick to reach between the bars and bring out each stone (steps 2, 3, and 4).

After a short time for consideration, he took the first stone and dropped it into the tube on the top of the clear plastic box with the large stick inside (step 5). It didn't open the door as he expected. He repeated that step with the second stone (step 6) into the same tube and box. Still, nothing happened. Once more, he dropped in the third stone (step 7), and this time the weight of the three stones together was enough to open the downward-hung door.

Now, 007 was able to reach inside and grab the longer stick. He walked over to the long, plastic box with the food inside and used the long stick to force out the tasty morsel (step 8). Success!

It took eight sequential steps to achieve. The researchers at Auckland claimed that this was the first known demonstration of spontaneous, sequential tool use by an animal other than a human. The creature who did it was a crow.

It is well known that animal intelligence is correlated with the size of the brain relative to total body mass. So, how does a crow compare to other animals? Scientists use a concept called the Encephalization Quotient (EQ) to explore this relationship. The higher the EQ, the more intelligent the animal. For example, homo sapiens has an EQ of 8.1. This compares to 5.9 for a bottlenose dolphin, 4.2 for a chimpanzee, and 4.1 for a crow. So, you can see that a crow is right up there in the hierarchy of intelligent animals. For comparative purposes, our beloved dogs, whom we believe are very bright, only have an EQ of about 1.4. Who would have thought that a crow would have the potential to be that much more intelligent than "man's best friend"?

One surprising capability demonstrated by wild crows is the ability to recognize and differentiate human faces. John Marzluff, a professor of wildlife science at the University of Washington, did some experiments that demonstrated this skill. There was a flock of crows in a park near his campus. The research team caught and banded seven of the crows while wearing a rubber mask of an ugly, ominous-looking caveman with a prominent brow and protruding teeth. For the next eight years, whenever someone dons that same mask and goes to the park, those crows would spot the masked intruder, become agitated, and begin scolding in alarm. Without the caveman mask, or even wearing a different control mask, the crows would not pay attention to those same people. And the crows developed this behavior in just one trial.

Even more interesting is that, somehow, the crows communicated with each other so that, over time, most of the flock developed the ability to recognize the "caveman" whenever he would appear, even in a crowd. Who would have guessed that crows were able to exchange such detailed information so well?

In another example of advanced intelligence, research scientists at the University of Tübingen in Germany found that crows can recognize the number of things in a set. They trained the crows to recognize groups of dots, from one to seven. Then they changed the sizes and arrangements of the dots, and the birds still recognized the number. Neuroscientist, Dr. Helen Ditz, said, "When a crow sees three points, seeds or even hunters, single nerve cells recognize the 'threeness' of the objects." And this is exactly how primate brains perform, she said, which is interesting since the brain structures of the two species are so different.

So, the next time you get upset with your local crows for making all that racket, or for opening your trash bags and making a mess, remember that they are among our most intelligent companions in the animal world and deserve a little respect, as hard as that might be.

Update of the article

Relative measure of crows' intelligence - Traditionally, brain size relative to body size, known as the encephalization quotient (EQ), has been used as a proxy for intelligence across species. However, this measure doesn't always paint a complete picture. Suzana Herculano-Houzel, a neuroscientist at Vanderbilt University, has pioneered a method called the ***isotropic fractionator*** *to accurately count the number of neurons in animal brains,* offering a different perspective on intelligence. Her work has revealed a surprising truth about bird brains. Despite their small size compared to mammals, they often *possess a high density of neurons, particularly in the forebrain regions associated with higher cognition.*

Herculano-Houzel's research has shown that corvids, especially crows and parrots, *have a significantly higher number of neurons packed into their smaller brains than many mammals of comparable size,* and in some cases, even rivaling the neural density of primates with much larger brains. This high density suggests a greater information-processing capacity per unit of brain mass. Therefore, while the EQ might suggest a certain level of intelligence based on relative size, Herculano-Houzel's work emphasizes the importance of the absolute number and density of neurons as a crucial factor in cognitive abilities. This helps explain how crows can perform cognitive feats on par with some primates despite having brains that are a fraction of the size, highlighting that "brainpower" is not solely determined by gross size but also by the efficient packing of neural units.

Blue Jays, the brainy, bully boys of Cape Ann

First published in the Gloucester Daily Times on September 5, 2024

The northern blue jay is the cheeky, charismatic character of Cape Ann's backyards. With their bold personalities, striking blue plumage, and distinctive head crests, these birds have captured the hearts, or sometimes the ire, of birdwatchers and casual observers alike. But beyond their attractive appearance lies a fascinating world of intelligence and complexity.

You cannot help but be impressed with the overwhelming presence of a blue jay at your backyard bird feeder. The relatively large, indigo creature swoops in and lands on the feeder while screaming loudly at the current residents, causing the smaller finches and sparrows to leave their perches immediately. Then the bully boy takes his time, dining alone to his heart's content; no wonder the jays have a reputation for being major-league intimidators.

Blue jays belong to the Corvidae family, a group of birds renowned for their intelligence. Their close relatives in our area include crows and ravens. Corvids have been closely studied and found to have significant problem-solving skills, the ability to use tools, and an impressive memory capacity. It is no surprise, then, that blue jays share many of these cognitive traits.

Historically, scientists believed that all birds were cognitively limited. How many times have you heard people use "bird brain" to insult an adversary? However, a professor at Vanderbilt University, Suzanna Herculano-Houzel, did some pioneering work on animal brain structure. She developed a unique methodology for accurately counting the number of neurons in the brain. When she began measuring the number of neurons in the brains of various animals, she found a strong correlation between that creature's cognitive abilities and the number of neurons in their cerebral cortex.

While studying avian intelligence, she demonstrated that the brains of corvids, such as blue jays and crows, are densely packed with interconnected neurons. She found that the corvid pallium, similar to the cerebral cortex in a mammalian brain, holds about as many neurons as you would expect to find in a primate's brain, like a chimpanzee or gorilla. Surprisingly, corvids exhibit intellectual capabilities that are comparable to those of apes.

But intelligence is not the only thing that makes blue jays fascinating. Due to their intellect, they are also highly social creatures, living in complex communities with intricate social structures and a prolonged developmental period before becoming independent. They communicate with each other through body language and various vocalizations and are known for forming strong bonds with their mates and offspring.

Blue jays, crows, and ravens are different than most other birds in that they mate for life. The male and female build their nest together. Once the eggs hatch, both parents take care of their nestlings with great cheer. The father is constantly engaged in finding and providing food, while the mother takes care of feeding her babies.

The tiny jays fledge, or grow out their flight feathers, twenty days after they hatch. Then they are ready to fly and leave the nest. Even at this point, the family of jays maintains its strong familial ties. They depart as a family, traveling their territory together. The parents accompany the young jays during their explorations, helping them search for food. The youngsters stay with their parents for about one to two months and are then ready to fly independently.

Professor Richard Racine, of Clark University in Worcester, found in his study of blue jays that the birds establish a dominance hierarchy in their social groups, which remains stable over time. Contributing to this stability, he found that, in later years, the group members' offspring continued to belong to the same social groups as their parents.

As another indicator of their intelligence, blue jays have demonstrated a sophisticated understanding of cause and effect. Experiments have shown that they can associate specific actions with particular outcomes. For instance, like their crow cousins, they can learn to manipulate objects to obtain food rewards, which requires high cognitive flexibility.

Blue jays are especially noted for their immense memory capacity. This helps them survive harsh winters in the Northeast. During the summer and fall, they hide a vast quantity of acorns and other nutrients in various locations around their territory. Then, when food is scarce in the winter, they return to those locations, retrieve the items, and keep themselves alive for another day.

But this blue jay behavior also plays a significant role in balancing the ecosystem. There are many more stored acorns in their territory than the birds can use. Those nuts that are not retrieved germinate and develop as saplings in these new locations. As a result, blue jays play a significant role in expanding the range of oak forests.

In yet another ecosystem contribution, blue jays are opportunistic feeders that consume a variety of insects, including caterpillars, beetles, and grasshoppers. This helps to control insect populations and protect trees from damage.

From watching your backyard bird feeder, it is clear that most bird species are dimorphic, meaning that the male and female look very different. The male is usually larger, more colorful, and more conspicuous than the female. But the corvids, including blue jays, are different. The appearance of male and female blue jays is remarkably similar in terms of coloring and size. With crows and ravens, the males and females are similarly black, with a slight size advantage for the males.

Blue jays are fascinating creatures with complex social structures and critical ecological roles. They are intelligent, adaptable, and crucial to the health of forest ecosystems. As we learn more about the complexity of blue jays, our appreciation for these remarkable birds continues to grow. They are not just colorful ornaments for our backyards but fascinating creatures that deserve our respect and admiration. So, the next time you spot a blue jay, take a moment to appreciate the incredible bird that it is. You might be surprised by just how much there is to admire.

Update of the article

Recent research on blue jay intelligence - Current studies highlight the remarkable intelligence of blue jays, positioning them firmly within the corvid family, which is renowned for its cognitive abilities.

For instance, research has demonstrated the blue jay's capacity for vocal mimicry, imitating hawk calls to warn other jays or potentially deceive other species. Captive studies have also revealed their aptitude for tool use, with blue jays observed using newspaper strips to retrieve food that is out of reach.

A significant recent development in blue jay intelligence research is the first documented observation of tool use in the wild. In October 2021, a blue jay in Ohio was photographed breaking off a piece of bark and using it to try to extract a spider from under another piece of bark on a tree branch. This groundbreaking observation, published in March 2025, provides the first scientific evidence of this behavior in their natural habitat, corroborating earlier findings from laboratory settings. This discovery challenges previous assumptions and opens new avenues for understanding the ecological contexts that might favor the development and use of tools in blue jays.

A 2024 study documented blue jays crafting tools from twigs, including straight probes for reaching into holes, hooked twigs for pulling insects, and leaf strips for scooping water. This behavior indicates advanced planning and an understanding of cause and effect. Moreover, these birds teach these skills to others, creating a culture of tool use within their communities.

Ongoing research aims to explore further the extent and complexity of their cognitive abilities in both controlled and natural environments, promising more insights into the intelligent lives of these vibrant birds.

Octopuses are an alien intelligence

The article was submitted to The Gloucester Daily Times but has not yet been published.

Living off our coast is an actual alien intelligence: the octopus. It has been several hundred million years since humans and octopuses had a common ancestor. As a result, we have evolved very differently and are as unlike physically as one can imagine.

In other words, they are entirely alien to us. Yet, we share one very important thing: a high degree of intelligence compared to most other animals.

The octopus is an invertebrate, meaning that it lacks a spinal column or, in this case, any bones at all. While the most common octopuses may reach about three feet long, the Giant Pacific Octopus grows up to sixteen feet.

Octopuses are both predators and prey. They are grand masters of camouflage, allowing them to hide from other predators easily. They do this by instantly changing their color, patterns of color, and texture to match their environment. How do they accomplish such a feat? Their whole body is covered with color-changing cells that they can control in fractions of a second. Not only color control, but also pattern control. Furthermore, their skin can be reformed to match the texture of objects in their environment, such as rocks or seaweed. Their camouflage skills, therefore, are unmatched in the animal world and far exceed those of the chameleon.

Perhaps most surprising about our physical differences is that they have three hearts, eight arms, and nine brains! *Unfortunately for them, however, their life spans are very short. Usually only one to three years.*

Dr. Jennifer Mather, a comparative psychologist at the University of Lethbridge in Alberta, Canada, has spent the last thirty-five years studying the intelligence of octopuses. She is part of a group of researchers who believe that these animals possess a surprising amount of brainpower. Her work focuses on understanding how octopuses' thinking processes evolved and comparing that with the evolution of cognition in vertebrates, such as humans, primates, and corvids (crows and ravens). Despite the extensive differences in the evolutionary paths of vertebrates and invertebrates, she has found that octopuses do indeed learn from their environment and apply that knowledge in ways remarkably similar to those of vertebrates.

Dr. Mather wrote in the Canadian Journal of Zoology that "A cephalopod (octopus) is first and foremost a learning animal, using the display system for deception, having spatial memory (knowing and understanding its environment), having personalities, and engaging in motor play. They represent an alternative model to the vertebrates for the evolution of complex brains and high intelligence, which has yet to be fully explored."

Aquariums worldwide that have kept octopuses as part of their exhibits have reported many stories of apparent intelligent behavior displayed by these creatures. They found that this complex predator can map its environment through observation and plan to achieve its objectives. Decades ago, the Brighton Aquarium in England gained firsthand knowledge of this capability.

They had an octopus on display in a tank. The wise cephalopod could get out of its tank at night when no one was around, crawl across the floor to the next tank, get inside, and eat a lumpfish. Then, the eight-armed predator escaped the lumpfish tank and wriggled back to its tank, where it was found the next morning looking completely innocent. It took more nights of similar behavior and the loss of several lumpfish before the aquarium staff identified the culprit.

Beyond the intelligence required, it may be surprising that an octopus could physically accomplish this feat. The reasons are due to its unique characteristics. It has eight strong arms with suction cups, giving it extraordinary mobility, even on land. Each arm has its own brain and can function independently. In addition, the cephalopod has a talent for squeezing itself through small openings, even though the holes may be much smaller than its body. It only needs an opening large enough to clear its beak and some cartilage in its head. The rest of their body can be compressed to allow them to squeeze through. This capability has earned them the nickname, the "Houdini of the animal kingdom".

Finally, octopuses normally breathe with gills, but when they are out of water and their skin is moist, they can absorb oxygen from the air for several minutes while they make their way across the floor. Or, between tidal pools as they do in the wild.

Many scientists worldwide, including Dr. Mather, have become intrigued by octopuses' behavior and comparative intelligence. One of these is Dr. Graziano Fiorito at the Dohrn Zoological Institute in Naples, Italy. He experimented to demonstrate how octopuses learn through observation.

Fiorito and his team placed two octopuses in adjacent tanks, separated by clear plexiglass so they could see each other. Each cephalopod had a clear plastic box in its tank with a delectable crab inside. The boxes had three openings with covers that required different types of motions to be released. One octopus had previously learned how to enter the crab container, while the other had never done so before.

The name of the inexperienced octopus was Candide, as in the novel by Voltaire. A local fisherman just caught the young cephalopod that morning. He could see the crab in the box, but didn't appear to know what to do about it. However, the octopus in the other tank had been in the laboratory for a few days and figured out how to open the box. He immediately went about his task, opened one of the covers, got into the container, and devoured the crab. Meanwhile, as he carefully observed his neighbor through the clear Plexiglas divider, Candide became increasingly excited. He was watching and learning what he needed to know to be successful.

At this point, the experimenters removed the crab containers from the tanks and placed a black plexiglass cover between them so the octopuses couldn't see each other. Then, they put the box with the crab back into Candide's tank.

Now, the young octopus knew precisely what to do. He immediately went to the same opening of the three in the box that the other octopus had used. He grasped that cover and made the same motion as he saw the other octopus make. He got into the container and had his meal.

The experimenters attempted to repeat this experiment with Candide several times. But each time, they rotated the box so that the chosen opening was in a different position. Candide went immediately to the correct opening, used the same motion, and successfully acquired his meal. This demonstrated to the experimenters that Candide had learned to imitate a movement and truly understood what had to be done.

In other experiments, octopuses learned to recognize symbols. They were taught that of three doors, the one with the symbol of a cross, wherever it was located, would lead to a tasty shrimp. With each trial, the symbols on the doors were rotated, but the cephalopod always went directly to the door with the cross, ignoring the doors with the triangle or the circle.

Like crows, octopuses have been observed by their aquarium keepers as being able to recognize the faces of people they either like or dislike. An experiment was done in 2010 by Dr. Roland Anderson at the Seattle Aquarium. He confirmed this capability. One researcher fed and played with their subjects, eight Giant Pacific Octopuses, while another, dressed in the same uniform, scratched them with an irritating object. The octopuses quickly recognized the two staff members by their faces. The cephalopods easily approached the staff member with the food, looking for their treats. But when they saw the researcher who had irritated them, they fled. The octopuses sometimes used their siphons to shoot water at them, showing their disdain for that person.

The octopus, a creature profoundly alien to us in its physical form and evolutionary trajectory, presents a compelling case for the independent evolution of high intelligence. With their complex nervous systems, problem-solving abilities, and capacity for observational learning, octopuses challenge our anthropocentric views on cognition. These remarkable invertebrates offer a unique window into the diverse pathways through which intelligence can emerge, underscoring the natural world's vast and still largely unexplored potential. As research unravels the mysteries of octopus behavior and cognition, we are left with a profound sense of awe and a deeper appreciation for our planet's astonishing diversity of life.

Update to the article

The evolution of Octopus intelligence — The remarkable intelligence of the octopus is a *fascinating case study in convergent evolution.*[6] This is an example of a lineage that is only very distantly related to vertebrates, *independently developing similar, complex cognitive abilities.*

Octopuses do not have a cerebral cortex in the same way that mammals do. The mammalian cerebral cortex is a layered structure located in the outer part of the brain, associated with higher-level cognitive functions. *Instead, the octopus' brain comprises multiple interconnected lobes (over 30 identified lobes*) that have evolved from ganglia, which are clusters of nerve cells common in mollusks. Some researchers have noted similarities in the neural architecture of certain octopus brain regions (like the vertical lobe, which is involved in visual memory) to vertebrates' structures, like a hippocampus.

Octopuses do not have as many brain neurons as the most intelligent non-human mammals, such as apes or dolphins. Instead, they have approximately 500 million neurons,

---

[6] **Convergent evolution** - when different lineages of organisms independently evolve similar traits because they adapt to similar environments or ecological niches. This process creates analogous structures that have similar form or function that were not present in the last common ancestor of those groups

comparable to those of a large dog. *The distribution of these neurons in the body is a key factor in their unique intelligence.*

Only about one-third of an octopus' neurons are located in their central brain (which is doughnut-shaped and surrounds their esophagus). *The remaining two-thirds are found in brain structures located in their eight arms.*

Several evolutionary pressures likely contributed to this seemingly bizarre development. As shell-less mollusks, octopuses faced significant predation pressure. Enhanced intelligence, coupled with their incredible camouflage abilities and flexible bodies, provided a *crucial advantage in evading predators and ambushing prey.* Their predatory lifestyle, which requires sophisticated hunting strategies such as stalking, luring, and problem-solving to access shelled prey, further drives *the need for cognitive complexity.*

Furthermore, the octopus's complex sensory world likely played a role in its evolution. Its sophisticated eyes, capable of detailed vision, and its highly sensitive suckers, which provide information through touch and taste, deliver a wealth of data to its nervous system. *Processing and responding to this intricate sensory input would have favored the evolution of more complex neural processing power.*

Since their lifespan is relatively short, the rapid life cycle may have *favored quick learning and adaptation, further selecting for efficient and powerful cognitive abilities within their unique biological framework.*

In summary, octopus intelligence stems from a relatively large number of neurons distributed throughout its body. Its unique nervous system has evolved independently of mammals, allowing for a different kind of intelligence well-suited to its environment and lifestyle.

Comparative intelligence: a dog vs. an octopus — The number of neurons in the brains of dogs and octopuses is comparable, so their intellectual potential should be similar. *However,* due to their distinct environments and evolutionary paths, they exhibit *very different forms of intelligence.*

For example, due to their co-evolution with humans and a long history of domestication, dogs excel at understanding human social signals and cues. They can interpret body language, emotions, and even specific commands. As mentioned in another article,[7] Some dogs have been shown to understand and obey more than one thousand commands, demonstrating a unique form of intelligence.

Dogs also have problem-solving skills. They can learn new tricks and solve puzzles, especially when food or rewards are involved. They are capable of some basic reasoning and can understand cause-and-effect relationships.

Octopuses, on the other hand, are exceptionally good at problem-solving and using tools. They are adept at escaping enclosures, opening jars, and using tools like coconut shells to create shelters. They can navigate complex environments and make decisions that show a high degree of cognitive flexibility.

Octopuses exhibit both short-term and long-term memory. They can learn by observation and have been shown to retain information about past experiences.

In summary, dogs are brilliant at social interactions, communication with humans, and emotional sensitivity. Their intelligence is more social and emotional.

On the other hand, Octopuses are incredibly intelligent in a more solitary, problem-solving, and tactile way. They excel in navigating physical environments and adapting to new situations. So, if you consider *problem-solving, learning by observation, and complex behaviors as the measure of intelligence, octopuses will take the lead.*

---

[7] See the chapter, "Cracking the Canine Communication Code".

Talking with the animals

First published in the Gloucester Daily Times on June 7, 2024

For centuries, the idea of conversing with animals in their languages has been the stuff of fairy tales and folklore. Dr. Doolittle's fantastic ability to chat with creatures great and small captured our imaginations. But science offered us little hope of ever accomplishing such a feat. However, the latest advances in digital bioacoustics technologies for gathering data and Artificial Intelligence (AI) for analyzing it are bringing us closer to that dream.

Until recently, scientists had only attempted to establish two-way communication with animals on human terms. One example was Koko, a lowland gorilla born in the San Francisco Zoo. She was “adopted” at an early age by Dr. Penny Patterson, President of the Gorilla Foundation. Koko received intensive training in American Sign Language (ASL) communication. Over her 46 years, she learned more than 1,000 words using signs, many of which she created herself. Dr. Patterson was convinced that Koko understood the language and was capable of clear, two-way communication with humans.

Other scientists now say that Patterson’s approach was based on human rather than animal terms. Instead of teaching a nonhuman to speak like us, we should have considered their inherent abilities to engage in complex communication and then used that understanding to establish two-way communication with the creatures on their own terms. This concept is driving the latest research in the field.

The first step in establishing two-way communication between humans and animals is to understand how animals communicate with each other. Accomplishing such a complex task requires considerable effort by the biologists who study them. These scientists must observe the animals in various situations and then associate the creatures' vocalizations with the corresponding behaviors or emotions.

Sophisticated digital bioacoustics technology is one key to gathering the necessary data. Tiny, lightweight microphones and digital recorders are attached to the backs of the creatures being studied, such as whales or bats. They can record continuously, 24/7, even in remote places that may be dark or have low light. And it is done without disrupting the ecosystem by introducing human observers.

The result of the new instrumentation is a deluge of data. Humans can't go through so many statistics and extract meaningful relationships. That is where Artificial Intelligence (AI) comes in. There are well-developed Natural Language Models that use algorithms to translate human languages from one to another. These same algorithms are now being employed to detect patterns in nonhuman communications.

Using these technologies, researchers in Israel have determined that bats have a much more complex and refined language than previously understood. Yossi Yovel of Tel Aviv University was the lead author of the study. For two and a half months, his team monitored the sounds and actions of two dozen Egyptian fruit bats. They recorded fifteen thousand calls, which were synced with videos of the animal's related behaviors. Then, using AI, the scientists identified the speaker, the intended recipient, the situation, and the behavior resulting from each call by correlating specific sounds with the connected social interactions captured on videos.

There were many startling findings. Like dolphins, bats have been found to have individual names, or "signature calls," that they use during their social interactions. Some bats argue over everyday things, such as which animal has access to specific food sources and who will sleep in which part of their habitat. In another type of situation, the researchers identified a male bat's call to a particular female concerning a mating advance. It is like a human calling to ask a partner out on a date.

Similar to human mothers who use "baby talk" to communicate with their babies, bats also speak to their offspring. While human mothers tend to raise the pitch of their voices when speaking to their babies, mother bats lower their pitch. Through these interactions, the young bats eventually acquire the ability to "speak" their first "words".

In addition to bats, other creatures have also been shown to possess advanced communication abilities. It was back in 1970 when we heard the first recordings of the complex songs of Humpback whales. Dr. Roger Payne, the founder of Ocean Alliance, now based in Gloucester, identified the songs from U.S. Navy recordings of Humpbacks near Bermuda. The songs made a big hit with people. This event spawned the public's interest in whales and launched the highly successful "Save the Whales" movement.

With the new technology available, scientists are now interested in gaining a deeper understanding of how whales communicate with each other. In 2020, the Project Cetacean Translation Initiative (CETI) was formed, with Roger Payne serving as an advisor. Its goal is to establish a two-way dialogue with Sperm whales ultimately. The reason for focusing on this species is that they live together in clans consisting of large numbers of closely related individuals and engage in extensive communication with one another.

Richard Moore, a philosopher of language at the University of Warwick, in England, says there is evidence that the Sperm whale's calls are well-formed sentences similar to our language. And that whales may be capable of communication that is "vastly more complex" than we have previously thought.

Another reason for focusing on Sperm whales is that they are the planet's largest-brained animals. They know each other through a clan-specific dialect of click sequences, known as codas. By studying these codas, scientists aim to gain insight into what the whales are communicating to one another.

The road to true whale communication is a long one. We may never fully understand their language in the same way we understand human languages. But when and if we are successful, the potential benefits are vast. Understanding whale speech, just as Dr. Doolittle did, could revolutionize our knowledge of the ocean environment. This will enable us to protect these magnificent creatures and their habitats better. And it would be the first step in an authentic, interspecies conversation, forever changing our relationship with the animal world. That would be a remarkable achievement.

Update of the Article

Other areas of research in animal communication - Animal communication systems can be incredibly complex and nuanced, often involving a combination of vocalizations, body language, scent marking, and other sensory cues. Deciphering these multifaceted signals is a significant challenge for scientists.

A complicating factor is that animals perceive the world differently from humans. For example, whales navigate using echolocation, a sensory experience humans cannot fully comprehend. This difference in perception makes understanding their communications particularly difficult.

Despite these obstacles, scientists and researchers have worked for many years trying to understand how animals connect within their species. Ethologists[8] observe animals in their natural environments while meticulously documenting their actions, interactions, and vocalizations.

More recently, to enhance their visual observations, researchers have been using video and audio recordings of the animals' behavior. These technologies generate vast amounts of data that must be analyzed for patterns and correlations. And scientists can be overwhelmed. So, in the last few years, they have turned to Artificial Intelligence to automate data analytics. This results in more efficient and effective investigations.

---

[8] An **ethologist** is a scientist who studies **animal behavior** in its natural environment. Ethology is a branch of zoology, and ethologists aim to understand the innate (instinctive) and learned behaviors of animals, as well as their evolutionary origins and ecological significance.

With these advances, there has been significant progress in animal communication research. Here are a few examples:

Honeybees: A classic example of this kind of data gathering and analysis was conducted by Karl von Frisch, a professor at the University of Munich. In his book, "On the Language of Bees", published in 1923, he demonstrated that the complex "waggle dance" of bees is used to communicate the location and distance of food sources to other bees in the hive. His work was well accepted, and he won the Nobel Prize in 1973.

African Elephants: Recent research indicates that African elephants have "names" for each other, addressing one another with specific rumbles. This was discovered by analyzing their calls and observing their responses.

Naked Mole Rats: Studies have shown that naked mole rat colonies have distinct "accents" in their soft chirping vocalizations, which help them identify colony members. These accents are learned early in life and are influenced by the colony's queen.

Vervet Monkeys: Research has documented that vervet monkeys have different alarm calls for different predators (e.g., eagles, leopards, snakes), and each call elicits a specific escape behavior from other monkeys.

Cats: Studies have identified a wide range of facial expressions in cats, suggesting they are more social and communicative through visual cues than previously thought.

The future of animal communication research likely lies in the continued integration of AI and machine learning with traditional ethological observation. As AI advances, there's speculation about more sophisticated interfaces that could potentially translate animal vocalizations or behaviors into human-understandable language, and vice versa.

A key challenge is to move beyond simply identifying signals to understanding the meaning and intent behind them. Do animals have referential communication (referring to specific objects or events)? Do they have a theory of mind (understanding what others know or intend)? These are complex questions that require further research.

In conclusion, the latest research in animal communication is leveraging advanced technologies to analyze animal signals in greater detail, while also continuing observational studies in natural settings. While the goal of having "conversations" with animals in the human sense remains far off, significant progress is being made in understanding the intricate ways animals communicate within their species and potentially paving the way for more meaningful interspecies interactions in the future.

Cracking the canine communication code

First published in the Gloucester Daily Times on July 25, 2024

For centuries, dog owners have sworn that their furry companions understood every word they spoke. We interpret the animal's excited barks and happy tail wags as indicating a deep level of communication between us and our "best friends". But can we truly converse with canines?

Like many dog owners, I believe that we can. We talk to them as we would to humans. Using the context of the situation, the canine seems to develop an understanding of what we mean. I measured the vocabulary of one of my dogs and found that the number of words and commands he understood approached one hundred.

I tried to determine if he could learn longer, more complicated commands. After training, I say, "First (do this) and then (get or do that)", and he understands what I want. For example, I may say, "First, upstairs, then biscuit". He immediately runs upstairs and waits at the top landing for his treat. Another example is when we are walking together down the stairs. I say "First (me), then (him)", and he carefully stays behind me all the way down.

Dr. John Pilley, a renowned behavioral psychologist and canine cognition researcher from Wofford College in South Carolina, demonstrated that dogs indeed excel at understanding human speech. He trained his dog, Chaser, a Border Collie, to have an impressive vocabulary. Over time, Chaser developed the ability to recognize and retrieve more than 1,000 different toys by name. Chaser's capacity to learn and remember the names of objects demonstrated an advanced level of comprehension.

Many dog cognition researchers, however, were skeptical of Pilley's results. Until recently, there was little scientific evidence that dogs understood the meaning of words the way we do. This changed when a group of scientists at Eötvös Loránd University in Budapest, Hungary, led by Dr. Marianna Boros, conducted an experiment that demonstrated this. The researchers' results were published last March in the peer-reviewed journal “Current Biology”.

EEG technology was used to monitor the dog’s brain activity during the experiment. The dog’s owner showed the animal an object that the dog knew. If the owner said, “ball,” but showed the dog a rope, the canine’s brain activity indicated his confusion at seeing the wrong object. This is the same EEG result that scientists find with human subjects. They concluded that dogs do indeed understand that specific words represent particular objects, just as we do. The scientists used several different dog breeds, including a Border Collie like Chaser, and found no difference in the various breeds’ abilities to understand our words.

But dogs' comprehension of what we want from them goes well beyond the words we speak. The animals are attuned to other forms of communication. Over thousands of years of domestication, dogs have become adept at reading physical and human cues.

Another important canine cognition researcher, Professor Alexandra Horowitz of Barnard College, part of Columbia University, studies the nonverbal communications of dogs. Her book "Inside of a Dog: What Dogs See, Smell, and Know" details her research on how dogs perceive the world. “While they may not grasp the intricacies of human grammar”, says Dr. Horowitz, “they have a remarkable ability to understand our tone of voice, body language, and even facial expressions.”

"Dogs are incredibly adept at reading human emotional cues," explains Dr. Horowitz. "A raised voice or furrowed brow can signal disapproval, while a high-pitched tone and a smile can communicate excitement. By noticing these subtle signals, dogs learn to anticipate our behavior and respond accordingly."

But communication is a two-way street. Our dogs have needs, and they learn to connect with us to tell us what they want. They cannot speak words but use a multimodal approach to accomplish their objectives, including visual cues, vocalizations, and touch.

Visual cues are a significant tool that dogs exploit to connect with us. They utilize their body language, facial expressions, tail movements, ear positions, and even eye gaze to convey their emotions and intentions.

When my dog wants to go out, he sits before me and stares into my eyes until he gets my attention. Guessing what he wants, I may say, "Do you want to go out?". He responds with excitement and anticipation, moving toward the door as he looks back to ensure I follow. He conveyed his desire for me to do something for him and showed me he understood my question.

Barks, whines, growls, and whimpers are all part of a dog's vocabulary to communicate even more directly with humans. The pitch, tone, and duration of these vocalizations can tell a lot about how your dog feels. When my dog is outside and wants to come in, he waits patiently by the door for me to come. He makes a low, plaintive bark if I do not appear in a few minutes. If that does not work, he turns to a louder, more determined howl.

Touch is also an essential communication tool for dogs. When my dog wants something from me, he may use his paw. The meaning he intends to convey is context-dependent. If he is anxious, he might let me know by giving me his paw without being asked and showing a wistful look on his face. If he wants to get my attention, he may hit my knee or arm with his paw and then use another signal to indicate his wants.

Understanding how dogs communicate opens doors to a deeper and more fulfilling relationship with our furry companions. Surprisingly, they know a lot of what we say to them. We can bridge the communication gap and create a truly special bond by paying attention to their body language, vocalizations, and touch cues. Remember, patience, consistency, and a love for all things "dog" are the keys to unlocking the secrets of canine communication.

Update of the article

Latest research on communicating with dogs - Interest in human-dog communication persists among university research scientists. Significant insights have been made into dogs' abilities to comprehend and respond to human cues. Notable studies include:

Dogs Using Soundboards to Communicate - A study led by Professor Federico Rossano at the University of California, San Diego, published in Scientific Reports in December 2024, examined 152 dogs trained to use soundboards with pre-recorded words. Over 21 months, data from more than 260,000 button presses revealed that dogs formed meaningful two-button combinations, such as "outside" + "potty," indicating deliberate communication rather than random behavior or mere imitation of their owners.

Dogs' Embodied Responses to Human Utterances - In November 2023, Mika Simonen from the University of Helsinki published research in the *Journal of Pragmatics*, exploring how dogs physically respond to human speech during interactive activities, such as playing fetch. The study found that dogs can distinguish between being directly addressed and not, responding to human commands with remarkable speed, often within 200 milliseconds. This suggests a sophisticated level of understanding of human communicative cues.

Dogs' Understanding of Human Roles in Cooperative Tasks - Research published in Scientific Reports in 2024 investigated dogs' comprehension of human partners' roles in cooperative tasks. The study demonstrated that dogs are more likely to coordinate with humans when such cooperation leads to better rewards, indicating an understanding of their partner's actions and the benefits of collaboration.

These studies collectively advance our understanding of canine cognition, highlighting dogs' complex abilities to process human language and actions.

The beneficent bats are in trouble

First published in the Gloucester Daily Times on August 26, 2019

If you sit out in your backyard on a warm summer night during the half hour after sunset, you may see, highlighted against the twilight sky, flying creatures darting and diving in, what appears to be a random pattern. Here in Massachusetts, those creatures are most likely the species of flying mammals known as Little Brown bats.

Over the years, I have had that experience many times during each summer, and I would look forward to it. Three or four bats would perform their aerobatics over our back lawn as they used their echolocation capabilities to identify the insects they crave and then dive on them to catch and consume their prey.

Sometime during the night, one or more of those bats would fly onto our upper, back porch and perch, upside down, in one corner. They would stay there, taking a rest, and, at some point during the night, leave the porch to return to their daytime lair, wherever that was. We knew this because, in the morning, we would find that corner of the porch filled with bat droppings, known as bat guano, but the bats themselves were nowhere to be seen. However, the cedar-shingled wall and ceiling of the porch showed scratch marks where the small mammals had held onto the wood with their sharp little claws.

Fortunately for us, the little guys were responsible for catching thousands of mosquitoes, moths, and other night-flying insects. According to the Massachusetts Audubon Society, each bat can, amazingly, catch 600 insects per hour. In fact, a Boston University scientist estimated that all the bats living within Route 128 in the Boston area consume 13 tons of insects each summer. If you are lucky enough to have them in your backyard, you should be grateful to them for helping you to be mosquito-free.

The most remarkable characteristic about bats is how they track and pursue their prey. Like our most advanced military aircraft, they have an amazing, radar-like capability called echolocation. The small creatures continuously emit high-frequency sounds (above human hearing) at up to 500 pulses per second. These pulses hit an object, like their insect prey, and bounce back to their sensitive ears, which are similar to sophisticated radar receivers. Their little computer-like brains can instantly interpret the return signal, which includes the position and speed information they need to locate their prey in the air. And then they swoop, with just the right amount of flight controls, to make the interception.

Unfortunately, many myths about bats persist, causing people to be unnecessarily afraid of them. For example, some believe that bats get tangled in women's hair. Not true, bats use their exceptional skills to keep well clear of people.

There is one thing, however, to be concerned about with bats. They can carry rabies. Less than 5% of them become infected, making it relatively rare. The bat's teeth are sharp, but tiny, so you may not even feel a bat bite. To avoid the risk of rabies, do not touch the little creatures. However, if you suspect you may have been bitten, it is crucial to capture the bat and have it evaluated carefully.

Last night, I was sitting on my porch looking for that familiar sight in the backyard. I waited and waited, but there was no sign of the bats. After forty years of seeing them in that location, I was disappointed that no one appeared.

Yet, I wasn't shocked. I had read that Little Brown bats were being affected by a strange, new disease. It's called White-nose Syndrome (WNS), and it has severely threatened their population. So, the absence of bats from my backyard caused me to assume that WNS had caught up with our local little hunters.

WNS first appeared in February 2006. Researchers were monitoring the bat population in a cave near Albany, New York, and noticed that many bats were exhibiting unusual behavior and dying. They were found to have a white, powdery substance on their faces, hence the name "White Nose." It turned out to be a newly discovered fungus which grows best in cool, high-humidity conditions, such as those found in the caves where the bats hibernate for the winter.

Many thousands of our Massachusetts Little Brown bats used to hibernate in the abandoned Emery mines located in Chester, Massachusetts. The Emery rock that came from the mines consists of the mineral coronadite, which makes an abrasive powder that's used for emery boards and emery cloth. Chester is only about fifty miles Southeast of Albany, New York, so it didn't take long for the fungus to appear there.

The aggressive fungus rouses the bats from their hibernation and increases their metabolism. This causes the bats to deplete the stored body fat that was intended to sustain them through the winter. So, they start to look for food outside the cave. But unfortunately, since it is winter and there are no active insects, the bats starve to death.

In the winter of 2007-08, there were estimated to be about ten thousand Little Brown bats in the Chester mine. By the end of the next winter, only fourteen bats were found alive. Since then, the number has increased to thirty individuals. So, there has been some recovery. And those bats are likely to have natural immunity to the fungus. But, since Little Brown bats only have one pup per year, it will take decades for the population to build back to previous levels.

According to the Massachusetts Audubon Society, the population of Little Brown bats has dwindled to less than 1% of what it once was in the state. As the fungus spreads across the country, approximately one million bats die each year. By 2030, scientists predict that the total population in the country will be reduced to 1%.

While Little Brown bats were the most numerous, there are several other species of bats in Massachusetts. Some of the others, like Large Brown bats, haven't been affected by the fungus. Since they don't hibernate in caves, they are not exposed to the deadly disease.

So, if you see a bat in your neighborhood, don't be afraid of it. In fact, be thankful that the little guy is still around, helping to keep your property pest-free.

Update of the article

The current status of White-Nose Syndrome on the Little Brown Bat population - The little brown bat population in Massachusetts has been severely impacted by white-nose syndrome (WNS), a fungal disease. In fact, the species is now listed as endangered under the Massachusetts Endangered Species Act.

Interestingly, some of the larger summer maternity colonies in Massachusetts are believed to be sustained by bats *that hibernate outside the state in sites less affected by WNS.* This offers a degree of hope that we will continue to see the little guys during the warmer months.

Research efforts are underway in Massachusetts and elsewhere to better understand the fungal disease and develop effective strategies for combating it. Some key areas of research include:

Bat Physiology and Immunity: Research aims to understand why some bats survive white-nose syndrome (WNS) while others succumb. This includes studying their immune responses, genetics, and physiological adaptations. Evidence suggests that some surviving bats develop a level of immunity.

Treatment Options: Researchers are exploring potential treatments, such as antifungal medications, to increase the survival rates of infected bats. Early treatment appears to be crucial.

Environmental Mitigation: Studies are investigating ways to modify hibernation environments. One approach is to maintain colder temperatures in the roosts. It is believed that this may slow fungal growth. Another is providing drier roosting sites in the summer through the use of bat boxes.

Vaccine Development: There is ongoing research into the development of a vaccine that could help bats survive WNS. A recent study showed that vaccination may reduce the impact of the disease.

While there is evidence of stabilization in some little brown bat populations in Massachusetts, the species remains endangered, and the long-term recovery is uncertain. Ongoing research provides valuable insights that can inform conservation strategies aimed at helping this once-abundant species persist in the face of this devastating disease.

Owls, denizens of the night

First published in the Gloucester Daily Times on December 28, 2023

Cape Ann is fortunate to have several species of owls living with us. They range in size from the small Eastern Screech Owl to the large Great Horned Owl. Occasionally, a Harry Potter-like Snowy Owl comes down from the Arctic and creates quite a stir on the Cape.

If you are lucky, you may encounter one of these creatures on your walks. One day, I was hiking in the woods when I looked up at a low-hanging branch of a tall, white pine tree. I was startled to see a large, Great Horned Owl up close and personal. If they have owlets in a nest nearby, a Great Horned Owl can become aggressive toward such a human intrusion. Fortunately for me, the big guy with sizable, nasty-looking talons just stared at me.

There is always a chance that you may see one in your backyard. A few weeks ago, my wife was working in our flower garden. She noticed some white spots on a Rhododendron bush, looked up, and saw a small owl roosting in the Maple tree just above the bush. It was an Eastern Screech Owl taking its daytime rest. The white spots were its droppings.

Through evolution, most owls have adapted exceptionally well to a nocturnal lifestyle. They possess exceptional senses of sight and hearing, which enable them to swoop down in pitch darkness and capture their prey silently. As a result, owls are truly the "denizens of the night".

The first thing you notice about an owl is its sizable, yellow eyes that are unlike those of any other bird. They are forward-facing with a focused gaze that gives the owl its appearance of being "wise". As a result of this configuration, owls have a wide range of "binocular" vision, or the ability to see with both eyes at the same time. This means that, like humans, they have the capacity for 3D sight and can visually judge the distance to objects of interest, such as a rodent on the forest floor.

Owls do not have eyeballs like ours. Instead, their eyes are in the form of an elongated tube, held in place by a bony structure. This arrangement does not allow the owl to move its eyes the way humans do. However, to compensate, they can tilt their head up to 270 degrees to the left or right from the forward-facing position.

Several features of owls' eyes contribute to their extraordinary night vision ability. First, their enormous size allows them to take in more ambient light. Their pupils can dilate to a very wide opening, allowing more light to enter the retina.

Yet another characteristic that enhances their superior night vision is that they have many times more "rods" in their eyes than humans do. The rods are cylindrically shaped, light-sensitive cells. In humans, they are concentrated on the outer edges of our retinas. Because of that, we see low light best when we do not look directly at the source. Not very convenient if you are a night hunter. But, with owls, their rods are clustered in the center of their visual field, allowing them to easily spot their prey while looking straight ahead.

Owls also have highly refined auditory systems to complement their extraordinary night vision. Like humans, their ears are located at the sides of their heads. They can hear a range of frequencies that are similar to those of humans. However, studies show that their awareness is focused on the frequencies of the sounds made by their prey moving in the underbrush.

Because, like humans, the owl's ears are on the sides of its head, there is a very slight time difference in the arrival at each ear of the sound of moving prey. The owl senses this tiny difference as the signals combine instantly in its brain. It knows where to look to spot its quarry, turns in that direction, and sees its target.

People have always thought of owls as being special. Even in Roman mythology, owls were associated with Minerva, the goddess of wisdom. They were supposed to presage events and were considered highly intelligent. As a result, astute, prudent people have always been considered to be "as wise as an owl".

Some birds, such as corvids (i.e., crows and ravens), and selected parrots, like the African Grey, are indeed highly intelligent. They can solve complex problems and easily learn to speak human language amazingly well. On the other hand, owls do not show such abilities.

Scientists who have studied owls found that the areas of their brains associated with sight and sound are larger and more complex than those of other birds, such as crows. It was observed that 75% of an owl's brain is devoted to its remarkable vision and hearing senses, which are critical to its survival. That does not leave much room for problem-solving and communication skills in the cortex. As a result, owls cannot think, learn, and improvise like crows. And, unlike crows, they do not form complex social structures and rarely communicate with others of their species.

So, the Romans were wrong about the owl being a wise bird. They are, however, beautiful creatures with a wide range of other skills. We can all appreciate it when we are fortunate enough to spot one in our backyards or while out walking in the woods.

Update of the article

Ongoing scientific research efforts on owls—There is considerable scientific interest and research on owls and their behaviors. Scientists want to understand their migration and movement ecology, foraging behavior, diet, and communication and social behaviors, so considerable efforts are underway.

Migration and Movement Ecology: Advanced GPS and radio telemetry tags are increasingly used to track owl movements, migration routes, habitat use, and dispersal patterns. This technology provides detailed information on their daily activities, foraging behavior, and long-distance journeys.

For example, Project SNOWstorm has been using GPS transmitters to track Snowy Owls, revealing their migration routes and wintering grounds. This endeavor is a collaborative effort involving dozens of scientists and organizations. It is crowdfunded"[9] and is one of the most significant ongoing research efforts focusing on snowy owls worldwide.

Foraging Behavior and Diet: Combining high-resolution GPS tags with video monitoring at nest sites allows researchers to precisely identify an owl's prey capture locations and the types of prey consumed. This helps the scientists understand the owls' hunting strategies and their impact on local ecosystems.

For example, the California Polytechnic State University's Humboldt Barn Owl Research Team uses GPS tags and video cameras to study barn owl hunting behavior in vineyards as a natural form of pest control.

Another interesting study used GPS tags and accelerometers on barn owls to understand their landing dynamics in relation to their hunting success. Professor Alexander Roulin conducted the study at the Université de Lausanne in Switzerland.

The Maine Owl Pellet Project is a collaboration between the University of New England, the Maine Department of Inland Fisheries and Wildlife, and the United States Fish and Wildlife Service. It engages community scientists in collecting and submitting owl pellets for analysis. Examining the bones and other remains in the pellets provides insights into the owls' diet and the distribution of their prey. They are especially concerned about the state-threatened northern bog lemming.

Communication and Social Behavior: Research continues to explore the complex vocalizations of owls, including territorial calls, courtship songs, and alarm calls. Studies

---

[9] **Crowdfunded** describes a project, venture, cause, or business that has received funding by raising relatively small amounts of money from a large number of people, typically through online platforms.

analyze the frequency, duration, and context of these calls to understand their meaning and function.

An example is the team at the University of Alberta. Professor Erin Bayne heads the team, which studies owl vocalizations using automated recording systems and software. The team's work focuses on detecting and analyzing owl calls to monitor populations and understand the impact of environmental factors, such as industrial noise, on owl communication.

The Cornell Laboratory of Ornithology has a vast archive of natural sounds at its Macaulay Library. The sounds include extensive recordings of various owl species. Scientists and researchers at the lab analyze these recordings to understand the nuances of owl vocalizations, their context, and the meaning behind them.

These are just some examples, and numerous other research projccts are underway worldwide, focusing on the diverse ecology and behavior of these fascinating birds. The findings from these studies are crucial for developing effective conservation strategies to protect owl populations and their habitats.

Resurrecting the woolly mammoth

Submitted to the Gloucester Daily Times but not yet published.

In a scientific endeavor that borders on the unbelievable, a group of researchers is embarking on a mission to bring back one of Earth's most iconic, extinct creatures: the woolly mammoth. The ambitious project, which aims to resurrect this prehistoric giant and create a dedicated habitat in its ancestral home of Siberia, is as audacious as it is controversial.

Woolly mammoths were large, elephant-like creatures from the Pleistocene epoch. They lived until about 10,000 years ago. Their habitat was in the world's coldest regions, such as the Siberian Tundra and the northern reaches of Alaska. As the climate warmed after the end of the last ice age, survival became more difficult for them. Exacerbating the situation, humans hunted the remaining mammoths until they were no more.

With rapidly advancing science and technology, however, we are on the verge of "de-extincting" the woolly mammoth as the first creature in human history to be brought back from the dead. The idea of resurrecting long-lost creatures originated in 1990 with the publication of Michael Crichton's novel "Jurassic Park", followed by a string of related movies directed by Spielberg. In those films, scientists used preserved dinosaur DNA to resurrect many of the prehistoric creatures. Unfortunately, some were extreme carnivores who escaped and wreaked havoc on the visitors of Jurassic Park.

The mammoths were massive animals, measuring approximately ten feet tall at the shoulder and weighing around six and a half tons. They were well-adapted for living in a frigid environment, cloaked in fur with an outer covering of deep brown, long guard hairs, and a shorter, yellowish-brown undercoat. Their ears and tails were relatively short compared to those of a contemporary elephant. This was an adaptation to minimize heat loss in freezing temperatures. The mammoths were also distinguished by having exceptionally long, curved tusks and four large molars for chewing their diet of grass. Like today's elephants, they used their tusks and trunks to manipulate objects, fight, and forage.

A company named Colossal Biosciences was founded in 2021 to resurrect particular extinct species, including mammoths. Their website says, "For Colossal, it is not just about making an organism that is, or resembles, an extinct species. It is about merging the past's biodiversity with the present's innovations to create a more sustainable future." The start-up founders were Professor George Church, a world-famous geneticist from Harvard University, and Ben Lamm, a serial entrepreneur who has started and successfully operated several venture capital-backed enterprises.

Colossal defines de-extinction as, "The process of generating an organism that both resembles and is genetically similar to an extinct species by resurrecting its lost lineage of core genes; engineering natural resistances; and enhancing adaptability that will allow it to thrive in today's environment of climate change, dwindling resources, disease, and human interference." So, they do not want to resurrect an extinct species but also genetically engineer it in a way that enhances its survival in today's World.

Ultimately, the company aims to repopulate the Arctic Tundra with the resurrected titans as a means to restore the ecosystem's balance. The trampling of the snow by the giant animals makes the ground colder, while creating open grasslands also induces the earth to become more frigid. The result will be to prevent the thawing of the permafrost, thus mitigating the effects of climate change.

To get started with their process, Colossal needed to have the woolly mammoth's genome. That is the complete set of mammoth genes present in one of their cells. Until recently, Colossal had only access to fragments of the creature's genome, as that was all that had been discovered. To overcome the problem, they were going to use the genome of an Asian elephant to complete the missing parts. This would have worked since the elephant has about 93% of the mammoth's genome, but the result would have only been a partial resurrection of the animal.

However, a significant discovery of a complete mammoth genome was announced a few months ago in a professional paper published in the peer-reviewed journal *Cell*. It is titled, "Three-dimensional genome architecture persists in a 52,000-year-old woolly mammoth skin sample". The animal was preserved when it died 52,000 years ago in Siberia because its body was essentially freeze-dried in the permafrost, thus preserving much of the creature. When the scientists examined the fossilized remains, they found the complete set of twenty-eight pairs of mammoth chromosomes preserved in the DNA fragments. So, Colossal will now be able to start its program with the complete mammoth genome.

The second requirement for achieving Colossal's goals is to have the necessary genetic engineering tools. The primary tool they will use is called CRISPR Cas9, a Nobel Prize-winning development from 2020. It allows scientists to edit a Genome sequence to achieve the desired changes relatively easily. For the resurrected mammoth to have a better chance at survival in today's world, Colossal may want to change or add some characteristics to the mammoth's DNA. This involves precisely cutting and replacing sections of their genetic material with certain corresponding, but more desirable, Asian Elephant DNA sequences. The modified mammoth cells will then be used to create an embryo that will be implanted into a surrogate elephant mother. Twenty-two months later, the world will see its first de-extinct woolly mammoth.

The same technology could also be used to de-extinct several other beneficial, lost creatures. One is the woolly rhinoceros, also a former inhabitant of the Arctic Tundra. Well-preserved carcasses have also been found in the permafrost, and their DNA should soon be readily available.

Whether or not we ever see living, breathing woolly mammoths or rhinoceros roaming the tundra again is still uncertain. However, the scientific advancements and dedicated efforts by Colossal and other research teams make it a possibility worth exploring. The result could be a boon for climate change mitigation. Pleistocene Park, a real-life location in Siberia with the same goal of preserving the permafrost, has indicated that it will welcome woolly mammoths when Colossal successfully brings them back. Unlike Jurassic Park, the creatures that inhabit Pleistocene Park will be herbivores rather than carnivores. So we can all feel a little safer when visiting the place.

Update of the article

Woolly Mammoth - Colossal Biosciences has made significant strides in its woolly mammoth project. Most recently, in March 2025, they announced the creation of "woolly mice". These are regular laboratory mice that have been genetically engineered using CRISPR-Cas9 technology to exhibit traits of woolly mammoths, specifically long, thick, and woolly hair, as well as altered fat metabolism associated with cold tolerance.

The birth of these woolly mice is considered a significant step and a validation of Colossal's gene-editing techniques for introducing complex combinations of traits into a target organism.

Colossal aims to produce mammoth-like Asian elephant embryos by next year (2026), with the first calves expected in 2028. Their ultimate goal is the functional de-extinction of the woolly mammoth, creating animals that can potentially be reintroduced to Arctic tundra ecosystems to help restore habitat and combat climate change.

Woolly Rhinoceros—While Colossal Biosciences lists the woolly rhinoceros as one of the species they aim to bring back, their research on this project is less advanced compared to that on the woolly mammoth and the dire wolf, discussed below.

As of now, there haven't been any major announcements or scientific publications detailing specific progress on the de-extinction of the woolly rhinoceros, similar to the "woolly mouse" milestone for the mammoth.

The challenges for the woolly rhinoceros project likely involve obtaining sufficient high-quality ancient DNA and understanding the complex genetic basis of its unique adaptations to the cold.

Dire Wolves - Colossal Biosciences recently announced a groundbreaking achievement: the birth of three dire wolf pups, marking what they claim is the world's first successful de-extinction of an animal species.

Scientists analyzed the complete genome of the dire wolf using DNA extracted from a 13,000-year-old tooth and a 72,000-year-old skull. They then made precise genomic edits to gray wolf cells to incorporate key genetic features of the dire wolf, focusing on traits like skull length, jaw power, fur thickness, and muscle mass.

The resulting embryos were transferred to surrogate domestic dogs for gestation. The three pups, two males named Romulus and Remus (born in October 2024 and named after the twin founders of Rome) and a female named Khaleesi (born in January 2025 and named after the dragon-flying Queen from The Game of Thrones, where Dire Wolves played a role), were created using gene-editing technology (CRISPR-cas9) on cells from living gray wolves.

These contemporary dire wolves are currently being cared for in a secure, 2,000-acre preserve certified by the American Humane Society. It is located in a secret location somewhere in the Northern United States. Colossal views this as a significant milestone, demonstrating the effectiveness of their de-extinction technology and its potential for broader conservation efforts.

In another recent experiment, Colossal successfully produced two litters of cloned red wolves, a critically endangered species. They used a novel, non-invasive blood cloning method, highlighting the potential applications of their research for extant species.

In summary, Colossal Biosciences is actively pursuing the de-extinction of the woolly mammoth and has achieved significant milestones with the birth of "woolly mice." They have also made a remarkable breakthrough with the birth of what they consider the first de-extinct species, the dire wolf.

Research on the woolly rhinoceros appears to be in an earlier stage compared to these other projects. The company's work is pushing the boundaries of genetic engineering and raising important scientific and ethical considerations about the future of conservation and our relationship with extinct species.

Black Bears set their sights on Cape Ann

First published in the Gloucester Daily Times on November 29, 2024

Surprisingly, black bears have started appearing along the coast of Massachusetts, raising questions among residents about their behavior and the potential long-term impacts on both wildlife and human populations. For decades, black bears have been known to inhabit the state's more forested and hilly regions, particularly in the Berkshires. However, recent sightings in coastal areas, including Cape Cod, the South Shore, and now the North Shore, have caught some residents off guard.

In Massachusetts, the large, brown muzzled mammals average 230 pounds for males and 140 pounds for females. In the 1970s, there were only about 100 black bears in Massachusetts. That number grew to 1,000 by the 1990s and is now estimated to be between 4,500 and 5,500, according to the Massachusetts Division of Fisheries and Wildlife (Mass Wildlife). Black bears continue to expand their range eastward. Their presence in central Massachusetts is now familiar. The growing bear population means bears and humans are more likely to encounter each other.

Black bears have become more common in the western suburbs. My son lives in the Town of Harvard, about 14 miles west of Concord. Last week, he was shocked when he looked out of his window and saw a black bear exploring his backyard. The bear got up on its hind legs to check a hummingbird feeder close to their back door. Then it turned and walked away, heading toward the woods and the next yard.

Closer to home, multiple recent bear sightings on the North Shore have occurred. MassBears is a project at Amherst College in collaboration with Mass Wildlife, whose objective is to record all the bear sightings in the state. Their website provides a sightings map with information on each reported encounter. In October 2020, a bear was sighted walking around in a wooded area near Cedar Swamp on the border of Manchester and Essex. Two bears were reported in separate sightings in Danvers in August and September of 2022. One of these creatures may have been the bear caught robbing a chicken coop in North Reading. Just this last August of 2024, one was spotted near Great Swamp Brook in Rowley. In Salem, also in August, a bear was seen eating trash in a residential neighborhood. And these are just the encounters that have been reported. There are likely many more than these.

The Black Bear Project Leader from Mass Wildlife, Dr. David Wattles, explains the phenomenon. “We’ve been tracking black bear populations as they spread from the Berkshires toward the state's more settled areas. Coastal sightings were unheard of just a few years ago, but the bears continue to push the boundaries of their range.”

So what’s driving black bears to explore areas traditionally considered out of their bounds, including coastal environments? Wattles points to a combination of factors, including the growing bear population in the state, the fragmentation of their traditional habitat, and the availability of food in the new areas.

Bears are omnivores, and their diets include everything from berries and acorns to insects, small mammals, and even carrion. As they move into more suburban and urban areas, they often find human-generated food to be very accessible and attractive.

Wattles says, “The bears are opportunistic feeders, and their movements are directed by following their food. Coastal regions, especially suburban neighborhoods with unsecured garbage cans, bird feeders, bee hives, and various crops in gardens and farms, can provide a wealth of resources for the bears.”

Wattles also notes that bears follow a natural pattern of population dispersal. While the females remain in their established habitats, the young males leave their mothers to seek new territories. "As the bear population grows, the young males move into previously unoccupied areas like the North Shore." For Gloucester and other Cape Ann communities, this means that these bears are beginning to investigate our region, and we may see more of them.

What can we expect as our interactions with bears increase? Black bears in wild areas are most active during the day. They are relatively shy and non-aggressive, but they are large animals with powerful claws and teeth, which can make them potentially dangerous. They have been known to kill goats and chickens. Unfortunately, such encounters usually result in the bear being eliminated by the animal owners or the police. To avoid such problems, Mass Wildlife strongly recommends using electric fencing to protect livestock. As a result, the bear is also protected.

Black bears very rarely hurt humans or pets, but it has happened. According to Dr. Lynn Rogers, founder of the North American Bear Center, the approximately 750,000 black bears living in North America will, on average, kill less than one human per year.

Dave Wattles is optimistic that, with proper precautions, conflicts can be minimized. "Bears are intelligent and adaptable animals, but also creatures of habit. If they don't find easy food sources, they're more likely to move on and avoid human-inhabited areas," he explains. So, if you don't want bears on your property, the most important thing you can do is to take down your bird feeders. Additionally, remove your beehives, electrify your chicken coop fence, and refrain from putting out your garbage for early collection.

Mass Wildlife relies on public reports to track problem bear activity and respond to emerging issues. The agency's approach to managing the expanding bear population involves a combination of public education, habitat management, and, when necessary, intervention to relocate problem bears.

The eastward movement of black bears onto Cape Ann is part of a broader trend that has been building for years. Thanks to conservation efforts and natural population growth, these animals are exploring new territories, bringing them closer to us here on Cape Ann. However, with guidance from experts like Dave Wattles and the proactive measures recommended by Mass Wildlife, Gloucester residents can prepare for this new reality and ensure that black bears remain a welcome part of the natural landscape.

Update of the article

Additional black bear sightings - After the article was published, I searched for additional sightings of black bears on or near Cape Ann and found none were reported. However, we can be certain that more black bears will migrate to our area in the years to come.

A black bear recently took the life of a man in Florida -As discussed in the article, black bears rarely hurt or kill humans. However, in early May 2025, Florida experienced its first recorded fatal black bear attack when 89-year-old Robert Markel and his dog were killed in his home in Jerome, a rural community southeast of Naples. The Florida Fish and Wildlife Conservation Commission (FWC) confirmed the incident, noting that Markel's remains were found approximately 100 yards from his residence. Evidence indicated a bear had entered his home and had a physical encounter with both him and his dog. The bear then dragged him from his home.

FWC officials euthanized three bears in the area as part of their conflict response protocol, and none tested positive for rabies. However, DNA testing linked a 263-pound male black bear to the attack, with the bear's genetic material found on Markel, his dog, and inside the home. A necropsy revealed partial human remains inside the bear, confirming its involvement.

This unprecedented incident has heightened concerns about human-bear interactions in Florida. The black bear population has grown to over 4,000, about the same as in Massachusetts. In response, the FWC is considering measures such as reinstating regulated bear hunts to manage the population and reduce potential conflicts.

While black bear attacks on humans are rare, this tragic event underscores the importance of Dr. Dave Wattles' advice on securing food and waste, especially in areas where bears are known to roam. The FWC and Mass Wildlife continue to advise residents on safety measures and remain vigilant in monitoring bear activity to prevent future incidents.

The majestic eagle, soaring again over Gloucester

First published in the Gloucester Daily Times on October 9, 2024

The bald eagle is one of the most iconic birds in North America. Known for its striking appearance, with its white head and tail contrasting against its dark brown body, this bird has come to symbolize freedom, strength, and resilience. As the national bird and symbol of the United States, the bald eagle has a rich history, fascinating biology, and a critical place in the ecosystems of North America. For many decades, we had not seen one soaring over Gloucester. But now they are back!

People have reported spotting them in various parts of the city. Then, last June, an article appeared in the Gloucester Daily Times about the state wildlife agency banding young eagles from a nest located somewhere in the city. Those eagles were noted as the first territorial pair recorded in Gloucester since early in the 20th century.

A few of my neighbors reported seeing eagles over Annisquam, so I started watching for the sight of one. I did see large, dark birds in the distance soaring over the beaches of Ipswich Bay. They were likely eagles, but I couldn't confirm it by spotting their distinctive white head or tail due to the distance.

But then last week, while having my morning coffee on our back porch, I looked up as an adult eagle flew into close view. It was low, about 100 yards in front of me, its bright white head glistening in the morning sun while it slowly beat its enormous wings to get over the trees on the other side of the clearing.

The bald eagle is found only in North America. Alaska is home to the largest population, with an estimated 30,000 individuals. These birds prefer habitats near large bodies of open water, such as lakes, rivers, coastal regions, and reservoirs, where their primary food source, fish, is abundant. The first wild, bald eagle I ever spotted in

Massachusetts was flying over the Merrimac River in Lawrence. That was about ten years ago.

Native American peoples considered the eagle a sacred creature. They regarded it as a messenger of the gods and used its feathers in their religious ceremonies. In 1782, early in the United States' history, Congress chose the bald eagle as the national emblem. They believed that the bird had a majestic presence and a reputation for strength, so it was a fitting reminder of some of our country's attributes.

Despite retaining its place of reverence in our country over the ensuing years, the bald eagle faced near extinction during the middle of the 20th century. This unfortunate prospect was the result of careless environmental management. Habitat destruction, hunting, and the indiscriminate use of pesticides, like DDT, decimated their population. As a result, by 1963, the number of breeding pairs in the lower 48 states had dwindled to just 417.

People everywhere started to recognize what was happening. In 1972, the detrimental effects of DDT were accepted as true, and it was banned from commercial use. The following year, the Endangered Species Act was passed by Congress and signed into law by President Nixon. These actions set the stage for the bald eagle's comeback. Today, the eagle has been removed from the endangered species list, with populations flourishing once again across the country. In Massachusetts, we now have over 90 territorial pairs, with one of the latest being the pair reported here in Gloucester.

The bald eagle belongs to the family Accipitridae, which also includes hawks, kites, and other raptors, such as eagles. It evolved from ancient birds of prey millions of years ago, and its evolutionary adaptations are key to its success as a top predator.

If you look at a bald eagle's face, it appears fierce. The bony ridge above its eyes causes this. The evolutionary purpose of this feature is to help shield the eagle's eyes from bright sunlight. This improves their chances of seeing their prey.

One of the bald eagle's most notable features is its eyesight. Eagles have bright, yellow eyes that are round and large relative to their heads. The back of their eye is flatter and bigger than humans, giving an eagle a much greater image size than we can see.

Humans have a special area on their retina called the fovea, where there is an enormous concentration of cone vision cells. In humans, the fovea has approximately 200,000 cones per millimeter. In the central fovea of an eagle, there are about five times more, or a million cones per millimeter.

As a result, an eagle's vision is enlarged and magnified compared to a human's and around four to five times sharper. These attributes allow them to spot prey from extreme distances. They can detect their target two miles away, whether it is a rabbit moving in the bush or a fish swimming just below the water's surface.

Anyone who lives near water knows about the effects of light's refraction on viewing objects below the water's surface. When you look at a fish under the water, it appears to be in a slightly different location than it is in reality. An eagle's eyes do not correct for this difference, so young eagles tend to miss catching their prey when they first start hunting. But they soon learn through trial and error to correct for the light's refraction, and then it is no longer a problem for them.

The bald eagle's journey from the brink of extinction to a thriving population is a testament to the power of conservation and the resilience of nature. As a symbol of North America's wild beauty and the values many hold dear, the bald eagle continues to inspire awe and respect. Whether soaring above the vast Alaskan wilderness or perched on a white pine tree in West Gloucester, this magnificent bird remains a powerful emblem of the continent's natural heritage.

Update of the article

Ongoing research projects for Bald Eagles – There are many ongoing research efforts aimed at improving the survivability of Bald Eagles. The overarching goals of current bald eagle research include:

Monitoring Population Health: Tracking population numbers, reproductive rates, and overall health to ensure the long-term stability of bald eagle populations.

Identifying and Mitigating Threats: Understanding and addressing factors that negatively impact bald eagles, such as lead poisoning, habitat loss, collisions, and entanglement.

Understanding Adaptation to Human Environments: Studying how bald eagles are adapting to living in closer proximity to human development to inform management strategies.

Informing Conservation and Management Strategies: Providing scientific data to guide conservation efforts, land management practices, and regulations aimed at protecting bald eagles and their habitats.

These studies are crucial for ensuring that bald eagles' remarkable recovery continues and that this iconic species remains a thriving part of North American ecosystems.

One example of this type of research pertains to a significant health issue affecting Bald Eagles. Cornell University, a center for ornithological research, has focused on lead poisoning as a long-term threat to eagles. They identified bald eagles as particularly susceptible due to their scavenging habits on the carcasses of animals shot with lead bullets and their proximity to the lead found in fishing tackle.

This activity is complicated by the eagle’s physiology, which allows them to mobilize lead quickly. Cornell’s work emphasizes that, while bald eagle populations have recovered significantly, lead poisoning might be slowing this recovery.

A significant study, "Demographic Implications of Lead Poisoning for Eagles Across North America," examined this issue and was published in the journal *Science* on February 18, 2022.

Data collected over eight years across thirty-eight states indicated that a significant percentage of bald eagles (47%) showed signs of chronic lead poisoning. This research underscores that even small lead fragments from ammunition can be lethal and accumulate in the eagles' systems.

Conclusion to Part II

In this exploration of animal behavior and intelligence, we have journeyed from black bears starting to visit Cape Ann to the enigmatic depths of octopus intelligence just off our coast. We have witnessed the resurgence of majestic bald eagles on Cape Ann and elsewhere in the Northeast, soared with the brainy blue jays, and pondered the "wicked smahts" of crows. We have decoded canine communication, delved into the lives of nocturnal owls, and considered the health challenges our bat populations face. We even entertained the idea of resurrecting mammoths, pushing the boundaries of scientific possibility.

A common thread emerges from these diverse narratives: the remarkable complexity and adaptability of the animal kingdom. Each species uniquely highlights a level of intelligence, social structure, and problem-solving ability that often surpasses our initial expectations. From the persistence and intelligence of black bears to the intricate communication and problem-solving abilities of crows, the animal world is a symphony of sentience, a testament to the power of evolution.

Moreover, these stories underscore the interconnectedness of all living things. The fate of the bald eagle is intertwined with the health of its ecosystem, the behavior of blue jays impacts the forest they inhabit, and the plight of bats reminds us of the delicate balance of nature.

As we continue to unravel the mysteries of animal behavior, we gain a deeper understanding of the natural world and a greater appreciation for our place within it. By recognizing the intelligence and sentience of other species, we can foster a sense of empathy and responsibility, inspiring us to become better stewards of our planet. The journey to decode animal behavior is far from over, but with each new discovery, we move closer to a world where we can truly hear and understand the voices of the animals with whom we share this Earth.

Part III

Seeking Non-Human Intelligence: The Fermi Paradox and Beyond

III. Seeking Non-Human Intelligence: The Fermi Paradox and Beyond

Introduction to Part III

The yearning to know if we are not alone in the vast cosmic ocean is as old as humanity's ability to ponder the stars. From ancient myths of celestial beings to modern science fiction's bustling, galactic empires, the question of extraterrestrial intelligence has persistently echoed through our collective consciousness. This section delves into the heart of that enduring inquiry, navigating the intriguing and often perplexing landscape where scientific possibilities intersect with the profound silence that we observe in the universe.

Our journey begins with the foundational efforts to actively listen for signals from other civilizations, a quest chronicled in the article, "The Search for Extraterrestrial Intelligence". We explore the early aspirations and technological hurdles of projects like SETI, highlighting the unwavering hope that a deliberate message might one day pierce the interstellar void.

However, this optimistic pursuit soon encounters a formidable intellectual challenge. The article "The Fermi Paradox: 'Where is everybody?'" grapples with this seemingly contradictory state of affairs. Given the sheer number of stars and potentially habitable planets in our galaxy, the paradox asks a stark question: Why haven't we detected any definitive signs of extraterrestrial civilizations? We will examine the various proposed solutions, ranging from the profoundly unsettling to the tantalizingly hopeful.

The narrative then turns towards more anecdotal yet undeniably compelling aspects of the search. The article, "We are not alone," explores a specific instance: the persuasive testimony before Congress of a government whistleblower and the surge of public and scientific interest it ignited. This chapter will delve into the evolving discourse surrounding unexplained aerial sightings and their potential connection, however tenuous, to the possibility of non-human intelligence.

Juxtaposing this is an article, "'If UFOs are real, why hasn't somebody found one of their ashtrays?" This chapter injects a dose of grounded skepticism, examining the lack of tangible, irrefutable physical evidence that would definitively prove extraterrestrial visitation. It probes the logical inconsistencies and evidentiary gaps that often plague claims of alien encounters.

Our exploration then ventures into the realm of myth and speculation surrounding clandestine government facilities in, "The Truth is Out There". This chapter dissects the enduring fascination with the secretive location of Area 51, separating fact from fiction and analyzing the cultural impact of the persistent rumors of hidden alien technology and encounters.

Returning to more official channels, "The Pentagon: UFOs are real, and very mysterious", examines the significant shift in governmental discourse regarding UAPs. We will analyze the implications of official reports acknowledging the existence of unexplained aerial phenomena that defy conventional explanation, and the renewed scientific and national security interest this has sparked.

Through these diverse perspectives, from the patient listening of SETI to the perplexing observations of UAPs and the enduring enigma of the Fermi Paradox, this section aims to provide a comprehensive overview of our ongoing quest to understand our place within the cosmos and the tantalizing possibility of encountering minds not our own. The silence of the universe may be profound, but the human desire to break it remains a powerful driving force.

The search for extraterrestrial intelligence

First published in the Gloucester Daily Times on September 19, 2015

Carl Sagan was a popular astronomer who wrote the novel "Contact." It was about the Earth's first interaction with a remote, alien civilization. The book was adapted into a movie in 1997, starring Jodie Foster. The result of this exposure to the public was that many people became aware of the possibility of intelligent life beyond Earth.

In fact, the search for extraterrestrial intelligence (SETI) has been a serious scientific investigation since 1961. That was the year that astronomer Frank Drake published his well-known paper, in which he provided an equation that estimated the number of active, communicative, extraterrestrial civilizations in our Milky Way galaxy.[10]

Harvard astronomer Howard Shapley did a similar estimate a few years earlier and speculated that there are 100 million *worlds in the universe* where life as we know it could have evolved. However, until the 1990s, there was no scientific evidence that planets outside our solar system (exoplanets) even existed.

The first exoplanet found around a star similar to our sun was in 1995. To date, astronomers have identified 1,977 exoplanets, with more than half having been discovered in the last five years. Once a new planet is found, other scientists study it to learn more about its characteristics, including whether our type of life could exist on it.

For example, in 2008, scientists analyzed the light coming from the exoplanet HD 189733 to determine the chemical components found in the planet's atmosphere. It was the first time an organic compound, methane, was identified on an exoplanet, along with water vapor and carbon dioxide.

---

[10] Drake's original estimate was optimistic, suggesting that there are between 10 and 10,000 advanced civilizations in the Milky Way. The variables were drawn from scientific knowledge at the time. The equation was a starting point for a conversation.

These discoveries indicate to scientists that life, as we know it, could exist on an exoplanet, as it has gases associated with life similar to those found on Earth. The probability of finding intelligent life, therefore, has increased.

There are numerous active SETI operations worldwide, searching for signs of intelligent civilizations beyond our own. Some of these organizations are supported by universities, others by governments, and still others by private donations. Two examples of the academic type are the programs at Princeton University and the University of California, Berkeley.

On the local scene, Harvard University has operated a SETI program since 1995. It is located at the university's Oak Ridge Observatory in the town of Harvard, Massachusetts. When I lived in that town, I had several opportunities to visit the facilities, see the large radio telescope with its super-computer banks, and interview some of the professional astronomers who worked there.

None of these programs has yet found evidence of intelligent life elsewhere in the galaxy. However, a recent development from NASA's Kepler space telescope mission has raised the hopes of some scientists.

Kepler, launched in March 2009, is a spacecraft designed to study thousands of stars in our Milky Way galaxy for signs of planets orbiting them. NASA scientists are particularly interested in identifying planets that may be suitable for life like Earth's.

The spacecraft detects the planets and gathers extensive information about them by recording the variability of the light coming from the star. If there is a planet orbiting the star, when the planet passes between the star and the Kepler spacecraft, it diminishes the star's light being recorded on Kepler's very sensitive sensors. The spacecraft then looks for the return of the same signal later, corresponding to a complete orbit of the potential planet around the star. In most cases, for the exoplanets found, the return signal is measured in days, with the number depending on the size of the particular planet's orbit. They have ranged from 2.2 days to 242 days.

Using this technique over the last few years, the Kepler spacecraft has found more than 1,000 planets orbiting 440 stars. Some of the planets were indeed found to be Earth-like in their characteristics. To be similar to Earth, the planet must be a solid, rocky mass and have an orbit around the star at the proper distance to maintain surface temperatures that allow water to exist in its liquid form.

But here is what has the astronomers excited about the potential for finding extraterrestrial intelligence. One particular star, designated KIC 8462852 (which is about 1,500 light-years from Earth), exhibited extremely strange results during Kepler's measurements. Most of the planets discovered by the Kepler spacecraft showed less than one percent diminishment of the star's light as the planet passed by. However, some of the signals seen over the last few years from KIC 8462852 diminished the starlight up to 22 percent, an enormous increase compared to the 1 percent diminishment of even a very large, Jupiter-size planet. The signals were also erratic in terms of their period.

Scientists considered many natural phenomena as explanations for the odd and unexpected results, but none seemed to explain the observations adequately. There are still a few such reasons to be explored. Still, astronomers are now seriously considering the remote possibility that the cause is a megastructure created by an intelligent, alien civilization. Such megastructures, possibly a super-sized solar array, have been theorized as a way advanced civilizations could obtain power by capturing some of the energy released by their star. They would be "mega" versions of the solar power satellites I discussed in a recent column.[11]

---

[11] See "Our Energy Future," published in the Gloucester Daily Times on October 15, 2015, and elsewhere in this book.

One of the SETI groups currently examining KIC 8462852 is the SETI Institute in Mountain View, California, a private, non-profit organization. They began their observations of KIC 8462852 on Oct. 16, 2015; They are using their Allen (Radio) Telescope Array (ATA) to search for non-natural radio signals coming from the vicinity of the star. The ATA operates independently and continuously. If a signal that it receives appears to be of intelligent origin, it is identified by the automated analysis system. The system immediately alerts the scientists to check the characteristics of the reception and confirm its origin.

In case the potential civilization around KIC 8462852 is not broadcasting on radio frequencies, the SETI Institute is coordinating with the brand-new Boquete Optical SETI Observatory in Panama. This facility, which opened last month, is searching the area around the star for brief but powerful laser pulses that form an intelligent transmission from an alien civilization.

We will soon learn the results of the SETI organization's examination of KIC 8462852. All scientists agree that the chance is remote that the erratic results of the Kepler spacecraft observations will be explained by the presence of alien-built megastructures around the star. However, until they identify a natural phenomenon that aligns with the data, they will continue to consider the alien megastructure concept a viable working hypothesis.

Update of the article

As of April 2025, 5,867 confirmed exoplanets reside in 4,377 planetary systems, 985 of which host more than one planet. Additionally, thousands of candidate exoplanets are awaiting confirmation.

The discovery of exoplanets has been a collaborative effort involving numerous astronomers, scientists, and space missions worldwide. Some key milestones and individuals involved in the early discoveries include:

- Aleksander Wolszczan and Dale Frail: In 1992, *they announced the discovery of the first confirmed exoplanets* orbiting a *pulsar* (PSR B1257+12). *These were not orbiting a Sun-like star.*
- Michel Mayor and Didier Queloz: In 1995, they *detected the first exoplanet orbiting a main-sequence, Sun-like star*, 51 Pegasi b. *This discovery is considered pivotal and earned them the Nobel Prize in Physics in 2019.*

Since these initial discoveries, numerous ground-based telescopes and space-based observatories have made significant contributions to the field. Some of the most prolific missions include:

- *Radial Velocity Teams*: Many astronomers have used the radial velocity or "wobble" method[12] to detect exoplanets by observing the slight back-and-forth motion of a star caused by the gravitational pull of an orbiting planet. This method can be used with land-based optical telescopes. The Harvard-Smithsonian team of astronomers was one of the first to identify such a system.
- *Kepler Space Telescope*: Responsible for discovering thousands of exoplanets using the transit method (detecting dips in a star's brightness as a planet passes in front of it).
- *Transiting Exoplanet Survey Satellite (TESS)*: Continuing Kepler's work, TESS surveys the entire sky to find transiting exoplanets orbiting nearby stars. As of March 22, 2025, TESS has identified *7,525 candidate exoplanets*, of which 618 have been

[12] The **radial velocity method**, also known as the **Doppler spectroscopy** or the **"wobble" method**, is an indirect technique used to detect exoplanets by observing the effect they have on their host star.

confirmed. Many more are expected to be confirmed as work continues.

Latest Developments:

The field of exoplanet research is constantly evolving, with new discoveries and advancements in detection and characterization techniques. Some recent developments and areas of focus include:

- *Confirmation of Habitable Zone Exoplanets*: In January 2025, researchers confirmed the existence of a super-Earth, HD 20794 d, orbiting within the habitable zone of a nearby Sun-like star just 20 light-years away. Its proximity makes it a key target for future atmospheric characterization missions searching for biosignatures.
- *Atmospheric Characterization*: The James Webb Space Telescope (JWST), launched December 25, 2021, is crucial in analyzing the atmospheres of exoplanets in unprecedented detail. Recent studies using JWST have revealed complex atmospheric compositions, including the detection of carbon dioxide in a gas giant (HR 8799) and the modeling of a potentially rocky exoplanet's hot, thick atmosphere (TOI-270 d).
- *Exocomets*: Astronomers have obtained new images revealing belts of comet-sized bodies around 74 nearby stars, providing insights into the formation of planetary systems.
- *Discoveries of Unusual Exoplanets*: Scientists continue to find bizarre and unexpected worlds. They include a "featherweight giant" planet with a "fluffy"[13] atmosphere. There were also planets with extreme temperature swings. For example, in March 2025,

[13] A *fluffy atmosphere* in the context of exoplanets refers to an atmosphere that is much more extended and less dense than what would be expected for a planet of its mass and size, especially when compared to the gas giants in our solar system like Jupiter.

NASA announced the confirmation of a "super-Earth" orbiting the star HD 20794. This planet, HD 20794 *d*, has an eccentric orbit that takes it in and out of the "*habitable zone*"[14]. As it approaches its star, it undergoes runaway heating, and as it moves farther away, it freezes any potential surface water. This planet cycles between these extreme temperatures roughly every 300 days.

- *Advancements in Detection Methods*: Researchers are refining existing detection methods and exploring new techniques, such as astrometry (measuring a star's position in the sky with high precision) and gravitational microlensing (observing the bending of light around a star and its planet).
- *Focus on Planetary System Architecture*: There is increasing interest in understanding how multiple planets in a system interact and how different planetary architectures form and evolve. Discoveries of multi-planet systems and circumbinary planets (a planet orbiting two stars) continue to challenge our understanding of planet formation.

Continuing the search for extraterrestrial life — According to the latest reports from April 2025, K2-18 b is the exoplanet on which scientists have found the strongest hints yet of the chemicals of life in its atmosphere.

Researchers using the James Webb Space Telescope detected the chemical fingerprints of *dimethyl sulfide (DMS)* and potentially *dimethyl disulfide (DMDS)* in the atmosphere of K2-18b. On Earth, these compounds are primarily produced by microbial life, such as marine phytoplankton.

K2-18b is located about 124 light-years away in the constellation Leo and orbits a red dwarf star *within its habitable zone*, meaning it's at a distance from its star where

---

[14] The ***habitable zone (HZ)*** of an exoplanet is the region around a star where the temperature on a planet's surface could potentially allow for the existence of *liquid water*, given sufficient atmospheric pressure.

liquid water could potentially exist. The planet is *also considered a "Hycean" world*, a type of exoplanet larger than Earth with a hydrogen-rich atmosphere and a substantial liquid water ocean.

While this discovery is considered the *most promising evidence of potential life beyond our solar system to date*, scientists emphasize that more research is needed to confirm these findings and rule out non-biological sources for these chemicals. While DMS and DMDS are known to be produced by living creatures on Earth, some other, non-life possible reactions can produce these chemicals. So, before we can declare that we have found life on an exoplanet, these possibilities will have to be dismissed.

The search for and study of exoplanets remains a vibrant and exciting field, promising further groundbreaking discoveries in the quest to understand our place in the universe and the potential for life beyond Earth.

The Fermi paradox- "Where is everybody?"

First published in the Gloucester Daily Times on July 19, 2018

Enrico Fermi was an Italian-American physicist and pioneer in nuclear physics. In 1942, he headed the team that designed and built the first power-producing nuclear reactor. He was also a key member of the Manhattan Project that created the atomic bomb.

After the war, Fermi continued to work at Los Alamos National Laboratory in New Mexico. In 1950, he was having lunch with some colleagues, including Edward Teller, the Father of the hydrogen bomb. The group discussed a New Yorker cartoon showing aliens emerging from a spaceship. Fermi had made some rough calculations on the probability of intelligent life in the universe. He concluded that intelligent beings were indeed probable and should have spread throughout the universe by this time. But "Where is everybody!" he exclaimed to his friends.

Fermi said that if aliens existed, they would have had billions of years to colonize the universe and should have left visible signs of their presence everywhere. But all observations of the cosmos indicate a "dead" universe. This has come to be known as the Fermi Paradox and has been a key point of discussion in astrobiology ever since.

Now, 68 years after Fermi made his statement, we are even more certain of its validity. Thanks to observations from land and space-based instruments like NASA's Kepler spacecraft, we know that, in our Milky Way galaxy alone, there are estimated to be over 100 billion exoplanets, many of which would be like Earth.

According to some analyses of the possibilities, using the Drake Equation,[15] thousands of these planets should have also produced intelligent life. Many of these civilizations

---

[15] Astronomer Frank Drake is a pioneer in the Search for Extraterrestrial Intelligence (SETI) and developed the Drake Equation to estimate the number of intelligent civilizations in the Milky Way Galaxy that can communicate with us.

would have been formed millions, or even billions, of years before us. It is easy to imagine that these beings would have created the means to travel to other stars. After all, it only took us 66 years to go from the Wright brothers at Kitty Hawk to Neil Armstrong stepping on the moon. Imagine how much space travel will improve over the next million years, a cosmologically short time!

If we look back at human history, expansion and colonization are prime directives for our species. It is highly probable that if an advanced civilization like ours had developed light-speed, space travel capability, its civilization would have exploded into the galaxy.

Yet again, there is no sign of them.

Over the years, many scientists have proposed explanations for Fermi's Paradox. Professor John Ball of the Harvard-Smithsonian Center for Astrophysics suggested a possible answer in 1973. He said the aliens are out there, and they are aware of us and other civilizations like ours. However, they have chosen to keep us in a type of preserve, allowing us to mature and evolve naturally. They will watch until they think we are ready to join the extraterrestrial community. And then they will contact us.

Another scientist who explained the Fermi paradox was Carl Sagan. In 1985, he commented on the lack of results from the projects engaged in the Search for Extraterrestrial Intelligence (SETI). Since 1961, the date of the first SETI experiment, most of these projects have utilized radio telescopes to attempt to detect signals from an alien civilization. And they have yet to hear the first whisper. Sagan said that we may not be listening for the Aliens in the way they are broadcasting. Their technologies are so advanced that their communication systems are unknown to us.

One of the key academic centers for considering the answer to the Fermi Paradox is the Future of Humanity Institute at Oxford University in England. Professor Robin Hansen, a member of the Institute, proposed the "Great Filter" theory in 1998 to explain the lack of evidence for alien civilizations.

Many other scientists have built upon his ideas and expanded them.

Essentially, Hansen stated that there are several stages of development that aliens on another planet must undergo to reach the point where their civilization can colonize the universe. Any one of these steps could be the Great Filter, where thousands of possible civilizations in our galaxy were stopped. So, they could never colonize the stars, which is why we don't see evidence of them.

The big question: Is that Great Filter behind us or ahead of us here on Earth? If it is behind us, that is good news. Earth has won the lottery. Everything necessary for intelligent life to develop and evolve came into place, even though all the probabilities were against it. Now, we will be the first to establish interstellar space travel and spread throughout the cosmos over the next billion years.

On the other hand, if the Great Filter is ahead of us, then there were probably thousands or even millions of civilizations before us, all of which hit the Great Filter and were destroyed before they could reach other stars.

How will we know which alternative is correct? The question remains unanswered if the planned NASA mission to Mars in the 2030s finds nothing. If, on the other hand, the mission discovers the *remnants of an ancient civilization*, that would be bad news indeed. It would confirm that the Great Filter is ahead of us, and as Robin Hanson says, "We're next!".

Update of the article

Solutions to the Fermi Paradox - The Fermi Paradox, which asks "Where is everybody?" given the presumed high probability of extraterrestrial civilizations arising in the vastness of the universe and over the fourteen billion years of its existence, remains *unanswered.* There is no single, universally accepted solution. *Instead, a wide range of scientists and philosophers have developed numerous potential explanations.* They are often referred to as *"the Fermi Paradox solutions."*

Some of the scientists are:

Robin Hanson, of George Mason University's Department of Economics and the Oxford Future of Humanity Institute, developed the concept of the *Great Filter.* He formalized the idea as a probabilistic framework. It posits that some step in the development from dead matter to advanced civilizations must be extremely improbable or fatal, explaining the silence in the universe. His famous paper on the subject was "The Great Filter—Are We Almost Past It?" (1998).

Nick Bostrom, University of Oxford, Future of Humanity Institute, conducted a deep philosophical analysis of existential risks and their implications for the Fermi Paradox. He argues that finding alien life could be bad news, as it would imply that the Great Filter lies ahead.

His famous paper on the subject is "Where Are They? Why I Hope the Search for Extraterrestrial Life Finds Nothing" (2008).

Anders Sandberg, of the University of Oxford's Future of Humanity Institute, models the settlement of the galaxy by exploring the limits of civilizational expansion and the feasibility of space colonization. His key paper on the subject is "That is not dead which can eternal lie: the *aestivation hypothesis*[16] for resolving Fermi's paradox."

Stuart Armstrong, University of Oxford, Future of Humanity Institute, proposed that advanced civilizations might "hibernate" to wait for a more incredible universe in which computation is more efficient, thereby explaining their current silence. He coauthored the Sandberg paper on the *aestivation hypothesis* listed above.

Stephen Webb, University of Portsmouth, is the author of "If the Universe Is Teeming with Aliens... Where Is Everybody?"

---

[16] The **Aestivation Hypothesis** is a proposed solution to the **Fermi Paradox** that suggests advanced civilizations may exist—but are currently *dormant,* or "aestivating," and waiting for better cosmic conditions before becoming active.

He catalogued and analyzed over seventy solutions, many of which are conceptual models for resolving the paradox. So, he is a synthesizer and analyst rather than an originator of a single dominant theory (*see the summary Categories below*).

David Brin is an Author formerly associated with the Jet Propulsion Laboratory of NASA who expanded on the idea that advanced civilizations might be deliberately hiding from or quarantining Earth as part of a non-interference directive.

Avi Loeb, Harvard University, proposed that objects like 'Oumuamua[17] might be evidence of extraterrestrial technology. He promotes the idea of gathering physical evidence through his Galileo Project, rather than relying on passive signal searching with programs like SETI.[18] While controversial, Loeb actively engages in empirical attempts to resolve the Fermi Paradox.

When looking at the totality of the work that all of these scientists have done, the solutions to the Fermi Paradox break down into four categories:

1. Aliens do not exist, so we will not see evidence of them.
2. They are here on Earth, but we cannot see them.
3. Aliens are present *in the Universe,* but communication with them is not possible.
4. They have colonized elsewhere in the Universe, but not here.

It's important to emphasize that these "solutions" are *hypotheses and speculations,* as we currently have no confirmed contact with extraterrestrial intelligence (ETI) or, more generally, non-human intelligence (NHI). Therefore, we cannot be certain which, if any, of these "solutions" of the Fermi Paradox are correct.

---

17 '**Oumuamua** is an object from outside our solar system that passed through briefly in 2017. It was the first known **interstellar object** to do so.

18 **SETI** is the Search for Extraterrestrial Intelligence.

Category 1: Aliens Are Not Here Because They Don't Exist (or Rarely Do)

- *The Rare Earth Hypothesis*: This suggests that the conditions necessary for the evolution of complex life and intelligent life capable of interstellar communication *are infrequent*. Enabling factors present in our case like a planet's size, orbit around its star, particular characteristics of gravity, the presence of a magnetic field, plate tectonics, having a large moon capable of stabilizing axial tilt, the presence of gas giants in specific orbits, and the particular history of our star system *might all be unique or extremely uncommon in coming together in other parts of the universe.*
- *The Great Filter*: This posits that *there is a significant evolutionary bottleneck or barrier that is very difficult for life to overcome.* This filter could be behind us (meaning we are fortunate to have passed it), currently in front of us (meaning our civilization is doomed to fail), or a combination of both.
    - *Filter Behind Us*: Examples include the abiogenesis of life itself, the development of complex cells, or the evolution of intelligence. If these steps are incredibly rare, then we might be among the first to make it this far.
    - *Filter Ahead of Us*: Examples include self-destruction through war, environmental catastrophe, or a technological singularity that leads to our demise before we achieve interstellar travel or communication.
- *Life is Common, but Intelligence is Rare*: Perhaps simple life is abundant, but the evolution of intelligence and technology capable of interstellar communication is a very improbable step.

Category 2: Aliens Are Here, But We Don't Recognize Them

- *They Are Observing Us (The Zoo Hypothesis):* Advanced civilizations might be aware of our existence but have chosen not to interfere or reveal themselves, perhaps to allow for our natural development or as part of a cosmic preserve.
- *They Are Too Advanced for Us to Recognize them*: Their technology and way of life might be so far beyond our comprehension that we would not recognize their presence or their signals.
- *They Are Here in a Form We Don't Understand*: Perhaps they exist in forms or dimensions that we are not equipped to perceive or detect.
- *We Are Not Looking in the Right Place or in the Right Way*: Our search methods might be too limited, focusing on specific types of signals or locations.

Category 3: Aliens Are Present in the Universe, But Communication is Difficult or Impossible

- *Vast Distances and the Speed of Light*: Interstellar distances are immense, making communication and travel incredibly challenging and time-consuming. Civilizations might exist, but are too far away for meaningful interaction.
- *Technological Bottlenecks*: Perhaps civilizations develop advanced technology but face limitations preventing interstellar communication or travel.
- *Civilizations are Self-Destructive*: Many civilizations might arise but destroy themselves before achieving interstellar capabilities or sustained communication.
- *They Don't Care to Communicate with Us*: Their priorities and interests might differ entirely from ours, and they might not see any value in contacting a relatively young and unsophisticated civilization like ours.
- *They Exist in Different Timeframes*: Civilizations might arise and fall on timescales that don't overlap with our own.

Category 4: Aliens Have Colonized Elsewhere in the Universe, But Not Here - for Specific Reasons

- *Resource Limitations*: Interstellar travel and colonization might be far more resource-intensive than we currently imagine, making it impractical or undesirable on a large scale.
- *They Have Different Priorities*: Their motivations for expansion or exploration might differ significantly from what we assume.
- *They Found Other, More Desirable Locations*: Perhaps other star systems offered more hospitable or resource-rich environments.
- *They Avoid Populated Systems*: Advanced civilizations might have a policy of avoiding contact with or colonizing systems that already harbor life.

In conclusion, the Fermi Paradox remains a mystery with numerous proposed solutions, yet no definitive answer has been found. Given the thought and energy that have gone into considering those solutions, the idea certainly inspires scientific inquiry and philosophical debate about our place in the cosmos.

We are not alone!

First published in the Gloucester Daily Times on August 15, 2023

The claim is shocking, to put it mildly. A highly credible whistleblower, Major David Grusch, USAF (retired), appeared under oath before the House Oversight Committee on July 26. According to Grusch, the U.S. government has retrieved many vehicles made by non-human intelligences over the last ninety years and has been studying those objects ever since. Astonishingly, the retrievals also included "non-human biologics"!

David Grusch was a high-level member of the intelligence community for fourteen years. He was part of the UFO-UAP investigation task force for the last four years. During the course of his work, he was informed by colleagues who worked on the retrieval programs about the recovered UFOs and the operations to reverse-engineer the technology for human use. When he tried to gain access to the programs for himself, his superiors rebuffed and punished him. Grusch launched a whistleblower complaint, which was investigated by the Inspector General (IG) of the Intelligence Community. The IG found that Grusch's allegations were "credible and urgent" and recommended that Congress investigate the matter. That was the impetus for the hearing on July 26th.

Previously, Grusch had testified in secret sessions before two Congressional committees where he provided evidence and more detail than he could in the open hearing. As a result, a bipartisan group of Congress, including Chuck Schumer, the Senate Majority Leader, has vowed to continue to investigate and get to the bottom of the claims. On July 13, the group of Senators introduced amendments to the annual National Defense Authorization Act that will require the Pentagon and civilian contractors to tell Congress everything they have done regarding "non-human technology".

When Grusch's claims are proven to be true, we will know for sure that we are not alone on this planet and in the Universe.

And "they", whoever they are, may have been with us for millennia, or longer.

History is filled with people reporting sightings of strange objects in the sky. In the biblical Book of Ezekiel, the prophet describes a flying chariot that he saw as "wheels within wheels, not standing upright, but laid down flat and moving as the appearance of a flash of lightning". This is very much as you would expect a person of that time to describe what people today call a flying saucer performing physics-defying movements.

In Nuremberg, Germany, on April 14, 1561, a printed news sheet reported that a large number of people in that city observed several glowing orbs maneuvering in the sky for about an hour. The orb is the most common shape of UFOs as reported by the Pentagon.

The current era of "flying saucer" reports began in 1947 with the highly publicized sighting by Kenneth Arnold, a private pilot. He was flying his plane near Mount Rainier in Washington State when he spotted a group of nine objects moving at a very high speed. He described them to reporters as being crescent-shaped and moving "like saucers skipping on water." The newspapers thereafter referred to the objects as flying saucers.

Over the next two decades, the U.S. Air Force launched several "studies" of the thousands of reports of Unidentified Flying Objects (the term coined by the Air Force). The Air Force ended their official program, Project Blue Book, in December 1969, saying that, if they had the time to gather the evidence, all of the UFO sightings could be explained as natural or manmade phenomena. Their conclusion was, " *There is nothing to see here.*"

The closing of Bluebook was the beginning of one of the greatest deceptions of the American public in history. This is because, all along, the government was likely in possession of a number of these objects and knew that they were not of human origin.

After 1969, the Air Force created an environment where a great stigma became attached to reporting something about

UFOs. They convinced us that anyone seeing a UFO was a kook or an oddball. And the tactic worked well for almost eighty years.

It wasn't until 2017, forty-eight years after the end of Project Blue Book, that the Pentagon finally acknowledged that the mysterious objects in the skies did, indeed, appear to be of non-human origin. They released gun camera footage from Navy F/A-18 fighters of three such encounters. The Pentagon, however, continued to claim that they had no idea whose vehicles they were or their intentions.

The next steps after the July 26th Congressional inquiry will include more hearings. These will likely feature additional whistleblowers whose names have been provided to Congress by Grusch. These witnesses will be able to testify with first-hand knowledge of the retrieved vehicles and the non-human biologics.

In the meantime, the legislation provided by the bipartisan Congressional group will require that the U.S. government and private contractors disclose that they possess recovered UFOs and "biological evidence of living or deceased non-human intelligence"! The amendment also requires the development of a "Controlled Disclosure Campaign Plan." This could be the path by which all of the secret UFO program data is ultimately released to the public. And then we will know with certainty that "we are not alone" in the Universe.

Update of the article

Interview with Jake Barber, second UAP whistleblower – In February 2025, News Nation, the TV news network, interviewed Jake Barber. The Air Force veteran claims to have been employed by the government as a helicopter pilot to work on retrieving crashed vehicles that he says were of non-human origin.

Barber recounted in a news interview his involvement as a helicopter pilot in the recovery of a UAP, which he described as a perfectly white, egg-shaped object. It was notable for its complete lack of any apparent propulsion system. He showed a night-time video of the object as it was retrieved. Intelligence community officials informed him at the time that this object was of "non-human origin."

During the retrieval operation, Barber claims to have undergone a profound personal experience. He describes suddenly feeling "possessed" by a loving, feminine energy or spirit connected to the object. This experience profoundly affected him, becoming a central part of his narrative and shaping his understanding of the nature of these recovered UAPs.

In addition to the white egg-shaped object, Barber described his involvement in the recovery of other UAPs that he characterized as flying discs with eight distinct sections when viewed from the helicopter above. He stated that multiple recoveries of these octagonal discs had occurred, some of which took place at night. The challenging terrain of the recovery sites sometimes necessitated daytime follow-up operations to salvage the rest of the material located on the site.

Barber has left his government affiliation and is currently involved in a similar, but private, project called *Skywatcher*. He co-founded the initiative with backing from venture capital. The group's primary objective is to *replicate the UAP work* that Barber claims he participated in during his time with the Pentagon.

Like the government operation, *Skywatcher* is actively and intentionally ***attracting*** *and observing UAPs* using methods that reportedly include meditative states, focused intention, and a "dog whistle" protocol, which Barber has vaguely alluded to as potentially involving psionics.[19] While the complete list of individuals involved in *Skywatcher* isn't

---

[19] *Psionics* is defined as the study of *the practical use of psychic powers*. This includes phenomena such as telepathy and psychokinesis, the interaction of material objects with psychic energy.

publicly detailed, Barber is a central figure who mentions a "team" working with him on this endeavor. *Skywatcher* is an ongoing project, with Barber making recent claims (as of January 2025) about their successes in "finding things that are not known objects". He suggests that their work is currently active and aimed at contributing to a broader public understanding and potential government disclosure of UAP phenomena in the near term.

It's important to note that Barber's claims are controversial and are being met with varying degrees of belief and skepticism within the UAP research community and the public. He and his team have created videos of their operations, including brief footage of object sightings, and uploaded them to YouTube. Barber has stated that he is more than willing to testify under oath before Congress to establish his credibility.

The current Pentagon UAP task force, the *All-Domain Anomaly Resolution Office* (AARO), is headed by Dr. John Kosloski. He has said they are interviewing whistleblowers, but he has yet to report what they have learned. There is some indication that the new Trump administration will release more definitive information on the government's involvement in UAPs. Many people are anxiously waiting.

The Pentagon: UFOs are real and very mysterious

First published in the Gloucester Daily Times on June 24, 2021

On June 25, 2021, or shortly thereafter, the Pentagon is going to release its Congressionally mandated report on Unidentified Aerial Phenomena (UAP). We have always referred to them as UFOs or Flying Saucers. Until now, anyone discussing the possibility of seeing such a thing was mocked and derided as eccentric or strange.

However, we now know that many people in the Military, using highly sophisticated technologies, have been seeing and recording these objects for years. The government still does not know what they are or where they come from, at least that is what they have been telling us.

In 2017, U.S. Navy videos taken by aircraft carrier-based fighters were released. They showed UAPs performing unbelievable maneuvers. At the time, many people thought the objects were tests of a secret U.S. project. According to a recent article in the New York Times, however, the Pentagon states in its upcoming report that the UAPs are not ours.

The Pentagon does not rule out the source of the objects as being Russia or China. We know both countries are experiencing similar incidents and have active UAP study programs. Furthermore, U.S. experts have gone on the record as saying there is no way that either of these countries could produce craft with the observed capabilities. These vehicles are not just one or two generations beyond our know-how, but probably hundreds or thousands of years ahead of us.

According to the New York Times article, the Pentagon will also say that it does not know if the UAPs are extraterrestrial in origin. They do not have any evidence to support this claim, but they cannot rule out the possibility.

The Navy and others have recorded some fascinating characteristics of the UAPs. Commander David Fravor, now a retired Navy fighter wing commander living in New Hampshire, had a close encounter in 2004. He and his wing mate observed what he called *a tic-tac-shaped object*. It was about the size of his F/A 18 fighter, which is 56 feet long. He has said on the record that he is unaware of any Earth-based technology that could duplicate its observed flight performance.

Another Navy video shows a rotating "gimbal" like craft. A third shows a sphere that quickly disappears into the ocean without being destroyed. Finally, a fourth object that a Navy Destroyer observed was shaped like a pyramid. In fact, there were several of them hovering above the ship that was operating off the U.S. coast. This form of UAP was, perhaps, the strangest of all.

All these objects have in common that they show no visible, aerodynamic surfaces, cockpits, or means of propulsion. There is no sound associated with them. They can go from low speed to thousands of miles per hour instantaneously. And at high speed, they can make a right-angle turn that creates astronomical g-forces that would kill any human occupant.

In many cases, there are *swarms* of UAPs in their operating areas. They range in number from ten to forty or fifty surrounding the nearby naval vessels. And they appear to be under intelligent control.

For many years, ordinary people have also been reporting sightings of UAPs, including right here on the North Shore. Most have been explained as airplanes, balloons, the planet Venus, the space station, or other natural phenomena, but some were likely Unidentified Aerial Phenomena (UAPs).

Now, given the confirmation that there really are unexplainable objects in the sky, you can help clear up the mystery. If you spot something unusual, take a close look at its details and try to record it with your iPhone's camera. Then, report it to one of the sites on the web.

I have always enjoyed looking up at the sky, watching meteor showers, comets, eclipses, and other natural phenomena. Once, however, I saw something I could not explain. And it happened right here in Gloucester.

My wife and I were standing on our back porch on a warm August night in 2016, about 10 p.m., looking out over Ipswich Bay. Off in the distance, I saw two bright lights that I thought were landing lights on a commercial passenger plane. I had seen that phenomenon many times before, so I was not surprised. But what caught my attention was that the lights were not white, the usual color, but orange. So, we continued to watch them.

The bright lights headed towards us and eventually passed overhead just in front of us. They were large, perfectly formed, bright orange spheres or orbs, which appeared slightly translucent with a more glowing, whitish center.

What happened next was even more surprising. We saw more of the objects coming in a line behind the first two, following the same course. Over the next twenty minutes, I counted twenty objects as they went by, all looking and behaving exactly the same as the first two.

I had never heard of orange orbs being reported as UAPs, so I did some online research. I found that it was a relatively common description, and there were hundreds of similar reports from around the world.

Several videos on YouTube showed exactly what we saw: a string of bright orange orbs moving along the same course.

The only explanation I could think of was that Chinese sky lanterns were released together and caught up in the wind. I reviewed videos of such lanterns to see if they resembled what we had seen. They were, however, *quite different.* Our UAPs were much larger, perfectly spherical, and much brighter than any candle-powered lantern. And when I *checked the recorded wind direction, it was from the Northeast, the direction in which the objects were heading. So, perhaps they were UAPs after all.*

The Pentagon UAP report is expected to be a turning point in the next few weeks. The government is acknowledging that UAPs are genuine and worthy of scientific investigation. NASA has recently announced that it has asked its scientists to study the phenomenon. In addition, it is almost certain that Congress will ensure that the military and intelligence communities conduct other urgent investigations. Hopefully, these studies will be made public at some point.

We are fortunate that the government is now recognizing the reality of UAPs. Knowing who created them and their intentions is vital to national security, and our future may depend on it.

Update of the article

The All-domain Anomaly Resolution Office (AARO), established by the Pentagon to investigate Unidentified Aerial Phenomena (UAPs), has released several videos on its official website between 2023 and 2025. These include:

Puerto Rico Object (2013): Footage captured by a U.S. Customs and Border Protection aircraft showing an object moving at high speed, splitting into two, and entering/exiting the water. *AARO assessed this as two balloons, flying near each other, not entering the water, but just appearing to do so.*

Mt. Etna Object (2018): Footage from an Air Force drone appears to show an object transiting a volcanic plume. AARO assessed this as a balloon distorted by atmospheric conditions.

Western U.S. Objects: AARO analyzed video and commercial flight data to conclude that these were three separate commercial aircraft.

Atmospheric Wake - South Asian Objects (2022): Multiple videos from MQ-9 drones showing objects with wakes. AARO assessed these as likely commercial aircraft with sensor artifacts.

Middle East Object (2022): A clip from an MQ-9 drone showing an unidentified object, though AARO found no anomalous behavior.

It's important to note that while the government has released these videos, the explanations provided by AARO often attribute the phenomena to known objects or atmospheric effects. However, some cases remain officially "unidentified."

AARO seems to be following the Project Blue Book playbook. *That is, focus on the objects that they can explain, and imply that, if they have enough time, the other sightings can also be explained.*

Plus ça change, plus c'est la même chose.

Strange metallic orb seen flying, landing, and then retrieved in Colombia - In March 2025, an unusual object was recovered near Buga, Colombia, a metallic sphere that has since sparked a wave of speculation and fascination. Measuring approximately 20 inches in diameter, the orb first drew attention due to its strange physical properties. Weighing in initially at 24 pounds, the sphere inexplicably dropped to 20 pounds shortly after its discovery.

Composed of dense, high-grade metal, its seamless surface—completely *lacking welds or joints*—immediately stood out to observers and scientists alike.

When the sphere was examined using X-ray imaging, the intrigue only deepened. Inside, analysts discovered a complex structure consisting of three concentric metal layers. Even more curious was the presence of 16 to 18 micro-spheres positioned around a central core, arranged in a deliberate and sophisticated design. *The object seemed unlike anything typically manufactured through known human technologies.*

Strange engravings etched into the outer shell of the orb added to the mystery. These markings *bore a resemblance to a wide range of ancient scripts,* including runes, Ogham, and even Mesopotamian characters. Some have interpreted these symbols as carrying cryptic messages—references to human consciousness, environmental balance, and forgotten histories—but *such interpretations remain speculative and controversial.*

Experts remain divided. Dr. Jose Luis Velazquez, a radiologist involved in the early examinations, *confirmed the lack of welds and noted the intricacy of the internal structures*, suggesting an artificial origin. Yet even he stopped short of attributing it to non-human sources.

On the other side, physicist Dr. Julia Mossbridge from the University of San Diego offered a grounded perspective. She proposed that the *sphere could very well be an elaborate art project or hoax, emphasizing the importance of rigorous, peer-reviewed scientific analysis before any definitive claims are made.*

Today, the object remains under investigation. No official declaration from Colombian authorities or international scientific organizations has confirmed what the sphere truly is—or isn't. Despite its puzzling features and the swirl of media interest, there is, at this time, *no conclusive evidence that the sphere is of non-human or extraterrestrial origin.* As with many such enigmas, only time, transparency, and meticulous research will reveal the truth.

If UFOs are real, why hasn't somebody found one of their ashtrays?

First published in the Gloucester Daily Times on August 14, 2021

The UFO era began in 1947 with the first widespread reports of strange objects in the sky. As the number of reports grew, the U.S. Air Force launched an investigation. In fact, between 1947 and 1969, the Air Force had several projects to study the phenomena, starting with Project Sign and ending with Project Blue Book.

The final report for Project Blue Book had two major conclusions. "There has been no evidence submitted to, or discovered by the Air Force, that sightings categorized as 'unidentified' represent technological developments or principles beyond the range of present-day scientific knowledge." And "there has been no evidence indicating the sightings categorized as 'unidentified' are extraterrestrial vehicles."

In short, Project Blue Book claimed to have solved the UFO mystery once and for all by chalking it up to natural phenomena. As a result of these public statements, the reporting or studying of UFOs was officially stigmatized. Professional observers such as pilots and radar operators feared reporting their sightings. And indeed, scientists avoided the topic. This remained the case for more than fifty years.

But on June 25, 2021, all of that changed. In a mandated report to Congress, the Pentagon finally, and reluctantly, admitted that UFOs, or Unidentified Aerial Phenomena (UAPs) as they are now called, are real and that they have no idea what they are, where they come from, or what their intentions are. This has sparked a new, widespread interest among the public, Congress, and the scientific community in studying the phenomenon and determining once and for all what these objects are and where they originate.

On July 30, 2021, about one month after the Pentagon UAP Report was issued, Professor Avi Loeb, a long-time member of the Astronomy Department of Harvard University, announced the formation of the Galileo Project. It is a privately funded research endeavor aimed at searching *for evidence of extraterrestrial technological artifacts (e.g., an alien ashtray). The idea is to apply systematic, scientific technology and research methods to investigate* UFOs and other phenomena that may have originated from extraterrestrial intelligence.

Professor Loeb is joined in the project by a team of scientists and engineers representing some of the world's top academic institutions. They include faculty members from Harvard, Princeton, Yale, UC Berkeley, and Cambridge in the UK, as well as Universities in Scotland, Sweden, Switzerland, and Israel. Having such a prestigious group work on studying UFOs would have been unthinkable before the Pentagon UAP Report.

Loeb believes that the best way to identify the nature of UAPs is to move away from relying on politicians or military personnel. Instead, professional scientists need to employ sophisticated scientific instruments and research methods to examine the phenomena in a completely transparent manner. They will only depend on the data they collect with their own technologies. This has never been done in any past investigation.

Avi Loeb first became interested in this subject a few years ago. It involved an astronomical event involving the visit of an unusual entity to our Solar System.

In October 2017, astronomer Robert Weryk, utilizing the Pan-STARRS telescope at the Haleakala Observatory in Hawaii, noticed a very strange object during his observations. He named it "' Oumuamua," which, in the Hawaiian language, means "scout" or "messenger."

Weryk and his associates noticed from the object's trajectory that, unlike most asteroids and comets, it originated from interstellar space. It had several other unusual characteristics. For example, the astronomers inferred from its reflection of the sun's light, as the object was spinning along its path, that its shape was more elongated, or flattened, than any known object from our Solar System.

Another intriguing observation involved the path it followed around the Solar System. The astronomers had calculated the expected trajectory using the Sun's gravitational force alone. But when they monitored its course, the object deviated from that path. Such a shift could be caused by the "rocket effect" that would be expected from "cometary" outgassing. But there was no cometary tail observed around 'Oumuamua, so outgassing could not be the cause. It had to be some other, unusual force.

These observations caused Professor Loeb to propose in a paper published in a professional astronomy journal that the object could be an alien probe. He also discussed this possibility in his later book on the object, **Extraterrestrial.**

He said in his article, "Oumuamua might be a defunct technological debris of equipment (from an alien civilization) that is not operational anymore." He theorized, "One possibility is that the object is a light-sail floating in interstellar space.... A light-sail is a sail pushed forward as it reflects light...." In fact, this was Loeb's explanation for why 'Oumuamua ***accelerated*** *out of our Solar System* without showing any cometary outgassing.

The Galileo Project has three main areas of research. First, it will obtain high-resolution, multi-detector, UAP Images to discover their nature. The scientists will search for UAPs with a network of mid-sized, high-resolution telescopes and detector arrays with suitable cameras and computer systems, distributed in select locations. To the best of our knowledge, neither the U.S. Air Force nor anyone else has ever conducted such an investigation before.

The second objective of the Galileo Project is to utilize existing astronomical surveys to identify and monitor interstellar visitors to our Solar System, similar to Oumuamua. The scientists also want to collaborate with people like Elon Musk and his SpaceX company to develop a launch-ready space mission that can "meet" an interstellar visitor to image and study it.

Third, they want to search for potential satellites, one meter in diameter or smaller, that they believe may be in orbit around the Earth and that were placed there by an extraterrestrial civilization to monitor Earth. If they exist, scientists believe they can find such objects using land-based telescopes equipped with advanced AI and fast-filtering methods.

This endeavor is exciting since it represents the first attempt by the science community to use our best resources and knowledge to study the UFO phenomenon. It is a privately funded, completely transparent effort that everyone will closely watch. After more than fifty years, we may finally get some answers.

Update of the article

The status of the Galileo Project, headed by Avi Loeb - The Galileo Project is currently in the active phase of data collection and analysis using its first observatory. The project's initial observatory is located at Harvard University. It is on the roof of the College's astronomical observatory in Cambridge, Massachusetts, and has been collecting data since 2023.

The Galileo observatory employs a sophisticated array of instruments, including an all-sky, infrared camera system with eight long-wave, infrared cameras. Over a five-month commissioning period (January to May 2024), the system recorded approximately 500,000 aerial objects. Artificial Intelligence was used to analyze this continuous data stream and identify outliers. About 16% of the recorded objects were flagged as potential anomalies and are undergoing manual review for further analysis.

The Galileo Project aims to establish two additional observatories in different locations in the Cambridge area. This will triple the data collection rate. The multiple stations will also enable the accurate measurement of distances to aerial objects through triangulation.

The project emphasizes a transparent, validated, and systematic scientific approach to investigating Unidentified Aerial Phenomena (UAP). It focuses on collecting high-quality data with precision-calibrated sensors and AI-driven detection methods. It does not engage in retroactive analysis of past events or anecdotal reports.

In summary, the Galileo Project has successfully deployed its initial instrumentation and is actively gathering and analyzing data. It is in the early stages of its long-term aerial survey and is working towards expanding its observational capabilities with the addition of new observatories.

The status of the search for objects originating outside of our solar system, like Oumuamua - The study of objects originating from outside our solar system has become a hot topic in astronomy, sparked by the discovery of 'Oumuamua in 2017. This object's unusual characteristics, such as its elongated shape and non-gravitational acceleration, puzzled scientists and ignited intense debate about its origins. While Oumuamua remains unique, it paved the way for future research and the search for more interstellar visitors. Another notable object, *Comet 2I/Borisov*, was discovered in 2019, confirming that these interstellar objects do indeed traverse our cosmic neighborhood.

Astronomers worldwide are actively involved in this search, utilizing various observatories and telescopes. Facilities like the Pan-STARRS telescope in Hawaii, which first spotted Oumuamua, continue to scan the skies for similar objects.

The upcoming Vera C. Rubin Observatory in Chile is expected to enhance detection capabilities due to its powerful survey capabilities significantly. Space-based telescopes, such as Hubble, also play a crucial role in characterizing these objects once they are discovered.

The scientific community is focused on observation and developing strategies for future encounters. Projects like *Project Lyra* explore the feasibility of sending spacecraft to intercept and study these interstellar objects up close. The ultimate goal is to understand the formation and composition of planetary systems beyond our own and perhaps even find evidence of extraterrestrial life or technology.

The truth is out there

First published in the Gloucester Daily Times on July 5, 2019.

Most people have heard of Area 51. The legend is that this is the place where the Federal Government takes retrieved Alien technology and bodies to be studied and dissected. It has been glorified in many movies, such as "Independence Day", and numerous TV shows, like "The X-Files".

One of the rumors is that captured UFOs are being reverse-engineered[20] there. The Alien technology is being investigated to determine exactly how it works. The objective would be to use that knowledge to create a similar craft for human use. Some people have claimed to have seen UFOs maneuvering around Area 51, presumably the captured alien craft that are being evaluated.

The truth is that the base in Area 51 has been, and remains, a strongly guarded Government facility used for the development and testing of top-secret U.S. aerospace technology since the late 1950s.

The location of Area 51 is in Nevada, about eighty miles northwest of Las Vegas. It is part of the vast amount of barren, moonscape-like federal land in that part of the state. On the maps, the land is divided into numbered Areas. Formerly, the most well-known Areas to the public were numbers 9 to 15. These were used as the Nuclear Test Site where, up to the 1960s, nuclear weapons were detonated in deep, underground bunkers. Area 51 is depicted on the maps, adjacent to the Test Site, and features a dry lakebed known as Groom Lake.

In the early 1950s, the CIA established its base in Area 51. It recognized the need for a way to gather aerial intelligence

---

[20] *Reverse engineering* (also known as backwards engineering or back engineering) is a process or method through which one attempts to understand, through deductive reasoning, how a previously made device, process, system, or piece of software accomplishes a task with very little (if any) insight into exactly how it does so.

over the Soviet Union. They wanted a plane that could cruise at 70,000 feet, out of reach, they thought, of Soviet defensive missiles. The Agency secretly commissioned Lockheed's, now famous, Skunk Works to design and build such a craft. The Skunk Works managers were responsible for finding an appropriate place on government-owned land to test their design. They, not the CIA, selected Groom Lake in Area 51 for its isolation and topographical suitability to conduct the tests. The U2's inaugural flight was in August 1957. This was the first mission at what was to become the Groom Lake base.

The second major project at Groom Lake was the successor to the U-2 spy plane. The U-2 was relatively slow and, as it turned out, not safe from Soviet missile technology, even at 70,000 feet.

Once again, the Lockheed Skunk Works was secretly commissioned to develop a solution. They developed the ultra-sleek A-12 Blackbird. A later, better-known model with the same "Blackbird" name was designated the SR-71.

The superfast, high-flying plane took over Soviet spy flights in 1966. The A-12 flew too high, 95,000 feet, and too fast, over Mach 3 or more than 2300 miles per hour, for the Soviet missile systems. So, no Blackbird was ever shot down. Later, however, the capabilities of advanced, satellite-based spy technology obviated the need for these aircraft, so after 1999, most of them were relegated to museums.

As the Groom Lake base in Area 51 expanded with other advanced development projects, the CIA lacked a sufficiently large staff and the necessary expertise to operate the site both physically and administratively. Instead, they contracted a Boston-area company to take responsibility for the base. That company was Edgerton, Germeshausen and Grier, Inc. (E.G. &G.).

E.G.&G. was formed after World War II by a group of scientists and engineers from M.I.T. who were heavily involved in the Manhattan Project. The firm became a major contractor to the Government for the nuclear weapons program. The CIA chose E.G. & G. for Area 51 because they were already operating the Nevada Test Site, adjacent to Area 51, and most of their employees were technologically sophisticated and held Top Secret clearances. E.G. & G set up their Special Projects Division, based in Las Vegas, to manage the CIA contract.

My second job as a young physicist was with E.G. & G. My responsibilities involved working in laboratories in the Boston area, where we simulated the effects of a nuclear explosion on various materials. The objective was to enhance the design of warheads on ICBMs to make them more resistant to enemy interference.

I never visited Area 51, but I did visit the Special Projects Division Headquarters and the Nevada Test Site, which bordered it. I had an opportunity to visit the Test Site as some of my colleagues were spending a few months there preparing for an underground nuclear test.

When I arrived in Las Vegas, I checked in with the Special Projects Division headquarters, which managed Area 51, to obtain an entrance badge for the Test Site. This was before Area 51 became known for popular stories about Alien technologies, so I was not particularly curious about the place. Furthermore, although I had a Top-Secret Q clearance, I did not need to know what was going on at the base. So, I didn't ask, and they didn't tell.

But in the last few years, many people who worked there have come out about what they did. One was Jules Kabat, an E.G.&G. Special Projects engineer employed at Area 51 at about the time of my visit. Many years later, he wrote about his two years at the secret base without violating his oath of confidentiality.

Kabat worked on the development of the A-12 Blackbird. His job, along with that of his colleagues, was to monitor the radar signatures of the Blackbirds as new materials and design modifications were introduced to enhance their stealth capabilities. His office faced the main runway at Groom Lake, the official name of the Area 51 facility, so that he could watch the Blackbirds' impressive takeoffs and landings. He did not report on seeing any signs of "alien technology" during his two years at Area 51.

Several other engineers and pilots also discussed their jobs at Groom Lake. Their activities were related to the significant developments in stealth aircraft. The F-117 stealth fighter was born here. As was the B-2 bomber.

I saw a B-2 flying over Boston during one of the Esplanade concerts on the Fourth of July. It was easy to understand how people reacted if they saw it during its development at Groom Lake. It is a large, black wing, flying low and quietly. It was unlike any other aircraft I had ever seen. If you didn't know better, it would have been easily mistaken for a UFO.

The discussion of reverse engineering Alien technology may also have its origins in the work conducted at Groom Lake. Captured Soviet aircraft were brought there for reverse engineering and testing. It was easy for conspiracy theorists to twist a leak about this type of activity at Area 51 to make it their own.

So, to this day, Area 51 remains an active Air Force base. In addition to the development of stealth aircraft, the first lethal drones were tested there back in the mid-sixties. More recent projects probably included the stealth helicopters that were used in the raid on Osama bin Laden's home in Pakistan.

There is no evidence that the base is involved in anything "alien." Yet, the rumors will likely continue as people spot test flights of the latest, most advanced, top-secret U.S. aerospace technologies. Most of these crafts will not resemble anything ever seen before, so it is easy to understand how rumors will continue.

Update of the article

Area 51 remains an active U.S. Air Force Installation -

E.G. & G., who provided technical and other support services to the base, was sold and resold. *Amentum*, a division of American Securities, now owns the Las Vegas operation that managed Area 51. This includes operating Janet Airlines, which transports workers to the base from Las Vegas' Harry Reid International Airport.

Janet Airlines departs from a dedicated, fenced terminal located at the northwest corner of the airport. The fleet, primarily Boeing 737-600s with a red cheatline, continues to transport military personnel, DoD civilians, and contractors to Area 51 (Groom Lake) and the Tonopah Test Range. Operations remain robust, with roughly 20 round trips per day between Las Vegas and top-secret test locations.

The Air Force administers Area 51 out of Edwards Air Force Base in southern California. Edwards is the home of the Air Force Test Center, where new military aircraft are tested and evaluated during development. The only confirmed use of the Groom Lake base is as a flight-testing facility for advanced and classified aircraft with the highest need for secrecy.

Due to its classified nature, the U.S. government does not disclose information about the specific research and development activities currently underway within the facility. However, it is widely believed to involve testing experimental aircraft, drones, and potentially advanced weapons systems.

If, as rumors have it, crash retrievals are stored at the base, perhaps we will learn more about them as more whistleblowers testify before Congress.

Conclusion to Part III

The conclusions drawn from these six articles paint a multifaceted and somewhat contradictory picture of humanity's quest for finding non-human intelligence and the enduring enigma of the Fermi Paradox. The established efforts of NASA and private organizations, such as the SETI Institute, as highlighted in “The Search for Extraterrestrial Intelligence,” underscore a persistent scientific commitment to listening for signals from other worlds.

The fact that, despite these ongoing endeavors, no confirmed extraterrestrial transmissions have been detected directly leads us to the core of “The Fermi Paradox: Where is everybody?” This article highlights the stark contrast between the theoretical abundance of potentially habitable exoplanets discovered by NASA and astronomers, some of which even exhibit promising biosignatures, and the complete absence of discernible alien civilizations. This paradox compels us to consider why, given the vastness and age of the universe, we appear to be alone.

The narrative takes a compelling turn with the articles addressing Unidentified Aerial Phenomena (UAPs). “The Pentagon: UFOs are real, and very mysterious” signifies a notable shift in official acknowledgment, admitting the existence of objects that defy conventional aerial technology and whose origins remain unknown. This aligns with the perspective suggested by “We are not alone,” hinting at the possibility that these enigmatic objects could indeed be the product of non-human intelligence, especially considering reports and historical accounts spanning millennia.

The implication, supported by government whistleblowers who claim to have retrieved and studied these craft, is that tangible evidence of extraterrestrial technology might already exist on Earth, albeit shrouded in secrecy. Programs like the Galileo Project and AARO, mentioned as initiatives to further understand these phenomena, represent a growing seriousness in investigating these unexplained aerial observations.

However, the article titled “If UFOs are real, why hasn’t somebody found one of their ashtrays?” injects a crucial element of skepticism and the demand for concrete proof. While acknowledging the intriguing nature of UAP sightings, it underscores the lack of definitive, publicly verifiable physical evidence that would unequivocally confirm an extraterrestrial origin. Despite numerous reported encounters over decades, the absence of known tangible artifacts raises valid questions about the nature and interpretation of these phenomena.

Finally, “Area 51: ‘The Truth is Out There’” delves into the realm of speculation and government secrecy surrounding this iconic location. While often associated with alien technology and encounters, the article navigates the blurry line between documented military research and unsubstantiated claims, highlighting the challenges in discerning fact from fiction in the pursuit of extraterrestrial evidence.

In conclusion, these articles collectively suggest a field grappling with both proactive scientific searching and reactive investigation of unexplained phenomena. While the Fermi Paradox continues to challenge our assumptions about the prevalence and detectability of alien life, the increasing official attention towards UAPs and the claims of retrieved non-human technology offer a tantalizing, albeit unconfirmed, alternative. The ongoing efforts of scientific institutions and government programs signal a renewed commitment to unraveling the mysteries of our cosmic neighborhood, even as the need for verifiable evidence remains paramount in this extraordinary quest.

Part IV

Our Dynamic Planet: Forces Shaping Earth and Beyond

IV. Our Dynamic Planet: Forces Shaping Earth and Beyond

Introduction to Part IV

The Earth and its forces are in constant motion, a dynamic interplay of geology, weather, and cosmic influences that shape the future of our planet and humanity. In this section, we delve into the cutting edge of scientific inquiry and into the visible and invisible forces that shape our world. From the ever-increasing shift of the Earth's magnetic field to the ever-present threats from space, each article examines the powerful phenomena that define the environment in which we live and the technological visions we are developing to meet future challenges.

Natural disasters, both rare and sudden, are a key focus. How vulnerable are we to a devastating tsunami off Cape Ann? How do rogue waves, once dismissed as sailors' tales, fit into our understanding of ocean science today?

Beyond Earth's waters, we look skyward: What would it take to protect humanity from an asteroid capable of triggering an extinction-level event? These topics remind us that, although our planet appears stable, it exists in a delicate balance, constantly subject to immense and sometimes unpredictable forces.

Then we dive into humanity's boldest ambitions, missions to Mars, the dream of practical fusion energy, and the quest for sustainable power with Space-Based Power Systems. These endeavors reflect scientific prowess and a larger vision for survival and advancement. They ask: How will we meet the energy demands of future generations? How can we ensure that our civilization can endure the inevitable cosmic and terrestrial dangers?

Finally, we tackle one of the most pressing and controversial issues of our time: climate change. Is it a matter of settled science, or has it taken on aspects of belief and ideology?

These essays form a sweeping survey of the forces shaping our past, present, and future, urging us to prepare wisely and dream boldly.

The Earth's magnetic field is on the move

First published in the Gloucester Daily Times on January 10, 2020.

Living on Cape Ann, there is a good chance that you are somehow involved with boating. If you travel for any distance in a boat, you need to understand the basic navigation principles. This includes using a magnetic compass and a marine chart to get you from point A to point B. When you follow your compass as you steer, you are relying on the fact that the Earth effectively has a giant bar magnet inside, generally aligned with its north-south rotational axis.

It may be surprising to some people that the Earth's bar magnet has its south pole located near the Geographic North Pole. Opposite poles attract, so that's why the north pole of your compass magnet points to it.

Mariners have depended on the Earth's magnetic field for centuries for navigation. In the nineteenth century, they realized that the magnetic field was moving and changing, affecting their long-range navigational accuracy. The government of Great Britain, a seafaring nation, established a program to monitor and track changes in the Earth's magnetic field, thereby improving its ability to predict its behavior more accurately.

Dr. Isaac Adams, the inventor of nickel-plating, lived in Annisquam during that period. In about 1898, he built a specially designed magnetic observatory in his backyard on Adams Hill. He used the instruments that he also designed to make very sophisticated measurements of the changes in the local Earth's magnetic field. The British used that information and the results from hundreds of other observatories worldwide to build a model for navigators that forecast how the magnetic field would further change over time. The mariners could then adjust their courses appropriately.

However, the Earth's magnetic field is much more useful to us than just for navigation. *In fact, without it, we wouldn't exist.*

Life, as we know it, developed on Earth only because our magnetic field was generated soon after the Earth was formed four billion years ago. The Earth's magnetosphere extends about 20,000 miles above the planet's surface.

There are two primary reasons the magnetic field is essential to our existence. First, it protects us humans from dangerous solar and cosmic particles that constantly bombard the Earth, and which would eventually kill us if they reached the surface. But, fortunately, the energetic, charged particles interact with the magnetic shield and are diverted away. Near the poles, where the magnetic field lines enter the Earth, we can see the results of these particles spiraling in. We call the phenomena the northern and southern lights.,

A second benefit of having charged particles diverted from the Earth's surface is that, over time, these particles would strip away our atmosphere. This is precisely what happened on the airless planet Mars, which doesn't have a magnetic field.

Why does Earth have a magnetic field and Mars does not? The difference is in their cores. At the center of both Earth and Mars are solid iron cores. But, in *Earth's case,* there is also *an outer core of liquid, molten iron alloy.* The sloshing around of this liquid metal ultimately forms our magnetic field.

As it turns out, the *Earth's North Magnetic Pole is not exactly lined up with its True North Pole.* That's why your marine charts have two compass roses. The inner one is the magnetic, while the outer one is the True North rose, with a star at the top.

The *North Magnetic Pole is the point on the globe where a compass needle points straight down.* The effective dipole magnet created in the Earth's core is tilted at about an 11° angle compared to the Earth's rotational axis. When the exact location of the North Magnetic Pole was first identified in 1831, it was located *at Cape Adelaide on the Boothia*

Peninsula in the northern part of the Canadian Arctic. This placed it several hundred miles from the geographic, or True North Pole, located in the middle of the Arctic Ocean. However, the magnetic pole hasn't remained at Cape Adelaide.

Geologists who study the history of the Earth's magnetic field have been tracking the movement of the Magnetic North Pole over time. When they first started tracing the pole's progress in the nineteenth century, it only moved at a rate of six miles per year. However, during the twentieth century, scientists were shocked to find that the pole movement had accelerated significantly, at a rate of about thirty miles per year. This was the fastest rate observed in the last four centuries. The pole has now wandered over 1,400 miles since 1831. Most recently, it is near Ellesmere Island in Nunavut, Canada. And it is headed toward Siberia in Russia.

For the average Cape Ann boater, this magnetic pole movement doesn't make that much of a difference to our relatively short-distance navigation. Whenever we receive a new set of charts, they include the latest information on the location of magnetic north, which is usually sufficient.

However, the rapidly changing field has an impact on large, commercial, ocean-going vessels and sophisticated military ships. To be accurate enough in their long-range navigation, they need to use the World Magnetic Model to adjust their courses. This is the current version of the model first developed by the British. It is now maintained by NOAA and updated at least every five years, or more frequently depending on the rate of pole movement.

One of the concerns regarding the faster movement of the magnetic pole is that it may be foreshadowing an even more dramatic change. Evidence of magnetization in rocks indicates that the magnetic poles have flipped approximately every 450,000 years. The North Pole has switched to the South Pole, and vice versa. *The last time was 780,000 years ago.* So, we could be due for another flip. No one really knows when it might occur.

The flip does not happen instantaneously. In fact, it could take about a thousand years. In the meantime, *the result would be a dramatic* ***decrease*** *in the strength of the Earth's magnetic field.*

Scientists first observed this decrease as the North Magnetic Pole began to move faster during the twentieth century. *They have measured a 10% decline in the overall magnetic field strength.* This could signal the beginning of a long-term switch. And this event *could lead to serious problems for us.*

The first thing that would affect us is the *impact of energetic, charged particles on our satellite systems.* The particles would destroy the onboard electronics, causing the satellites to malfunction. In the later stages of the pole switch, when the magnetic field strength is reduced to a small fraction of its current value, *the impact of solar and cosmic particles on humans becomes deadly.*

The good news is that the process of flipping is so slow that we would have time to prepare for those changes. Shielding would be provided on the satellites to protect the electronics. The effect of the particles on humans depends on the dosage or the level of exposure. As the problem of decreasing magnetic field strength worsens, humans will develop shielding solutions for their structures to protect the species.

Since the dawn of creation, humans have valued the Earth's magnetic field. Scientists are now paying close attention to how it is changing, and you can rest assured that they will direct us to make the necessary adjustments to deal with those fluctuations as required.

Update of the article

The shifting magnetic north pole—Over the past century, it has drifted away from its historical position near Canada toward Siberia. Before the year 2000, it moved six to nine miles per year. Then, in the early 2000s, it suddenly accelerated to 34 miles per year. By January 2020, the pole was located near Ellesmere Island in Nunavut, Canada.

Since then, however, *the pole's movement has slowed to about 15.5 miles per year in a south-southeast direction.* Starting in 2020, it moved another 77 miles and is now located northeast of Siberia.

This changing shift is due to the *complex dynamo theory*[21] *processes* within Earth's core, where the motion of molten iron generates the Earth's magnetic field. The pole's location and movement are not perfectly predictable but are tracked using instruments like the World Magnetic Model (WMM)[22].

2025 is a great year to see the Northern Lights - The aurora borealis, also known as the Northern Lights, is a breathtaking celestial phenomenon caused by the interaction of charged particles from the Sun with Earth's atmosphere, guided by the Earth's magnetic field. Most of the time, they are only visible at the highest latitudes, near the Arctic Circle.

The Sun constantly emits a stream of particles, known as the solar wind. When particularly energetic bursts of solar wind, often associated with solar flares and *coronal mass ejections* (CMEs), reach Earth, some of the charged particles are funneled in a spiraling motion along the planet's magnetic field lines towards the polar regions.

As these high-speed particles collide with atoms and molecules in Earth's upper atmosphere, they transfer energy, exciting the atmospheric gases. These excited gases then release this energy in the form of light, *creating the vibrant and dynamic displays of the aurora in various colors, such as*

---

[21] **Dynamo theory** explains how magnetic fields are generated and maintained in celestial bodies like Earth and stars. It posits that a dynamo, an electric generator, operates within a conductive, moving fluid (like Earth's liquid iron outer core) to produce and sustain a magnetic field. This process involves electromagnetic induction, where the *motion of the fluid* induces electric currents, and these currents, in turn, *generate and maintain a magnetic field.*

[22] **The World Magnetic Model** is a spherical, harmonic *model of the Earth's main magnetic field* and its slow temporal change. The WMM is key to global navigation, ensuring that technological systems relying on the Earth's magnetic field operate correctly.

green (from oxygen), pink and purple (from nitrogen), and sometimes red (from higher-altitude oxygen).

The Sun undergoes an approximately 11-year activity cycle, characterized by periods of high activity, known as solar maximum, and periods of low activity, referred to as *solar minimum.*

During the solar maximum, the Sun exhibits increased sunspots, solar flares, and coronal mass ejections (CMEs). These more frequent and intense solar events send more charged particles towards Earth, *leading to more frequent and spectacular auroral displays.*

According to NASA and other space weather agencies, the current solar cycle, *Solar Cycle 25, is predicted to reach its maximum activity in 2025.* The overall heightened solar activity, with increased CMEs, has dramatically increased the number of Northern Lights, making them visible on the North Shore of Massachusetts. These larger *solar flares are expected to continue through 2025 and potentially into 2026.*

This elevated activity means a higher probability of intense geomagnetic storms, which *expand the auroral oval, making the Northern Lights visible at lower latitudes than usual.* Therefore, 2025 is a prime year for witnessing the aurora borealis on Cape Ann, offering potentially brighter, more frequent, and more widespread displays.

Exploring the science of rogue waves

First published in the Gloucester Daily Times on April 23, 2018

Philip Weld, the former owner/publisher of the Gloucester Daily Times, lived on Cape Ann for many years. Late in life, he became interested in a dangerous form of sailboat racing: sailing solo across the Atlantic. He had a large trimaran built and named it "Gulf Streamer". He planned to enter the solo, cross-Atlantic race sponsored by the London Observer, an English newspaper.

In 1976, while on his way to England to join the start of the race, Weld's vessel was hit by a huge, "rogue wave" north of Bermuda. This caused his vessel to capsize. Once a trimaran turns turtle, it does not recover as the typical ocean-going yacht with a large, self-righting keel. Weld and his companion, Bill Stephens, had to cut their way through the bottom and crawl out onto the tri-hull. The exhausted mariners clung to the wet, slippery bottom for five days before a British container ship rescued them.

So, what is a "rogue wave" that could cause such a calamity? Although there were many anecdotal reports of massive, freakish waves before Weld had his encounter, experts thought that rogues were a myth. Scientists didn't understand how such waves could emerge from the storm systems that they studied, and no such wave had ever been measured reliably.

But in January 1995, on the Draupner oil-drilling platform in the stormy North Sea, a wave was digitally recorded with a maximum height of 84 feet, while the average waves in the area were a fraction of that size—a true rogue wave. *This changed everything.*

Scientists were now excited, as there was good evidence of the phenomenon. Numerous studies, both theoretical and experimental, have been conducted worldwide over the past twenty years. As a result, a much better understanding of the singularity has been developed. However, there is still a long way to go before useful predictions about the presence of rogues can be made.

One interesting finding in the research to date is that rogue waves are not as rare as once thought. The European Space Agency has been researching using radar images of the world's oceans to identify rogue waves. Dr. Wolfgang Rosenthal, a German scientist, said of the results, "They (rogue waves) are more frequent than we expected." He estimates that ten rogues are "churning through the world's oceans at any given moment."

Compared to the ambient waves in the area, a rogue is very large, unexpected, and appears suddenly from any direction. Such phenomena can be hazardous, even for major ships. The reason is that the rogue can impact an unsuspecting vessel with tremendous, breaking force.

For example, in a huge storm, like the "Perfect Storm" of 1991, a NOAA buoy east of Cape Cod reported maximum sustained winds of 56 mph. It also recorded average waves of 39 feet in height. These waves have a breaking pressure of 8.5 pounds per square inch (psi) on a ship's hull. Modern ships are designed to withstand a breaking wave with 21 psi, so normally, the waves in the Perfect Storm would not have been a problem. However, a large rogue wave created in such a storm can be 80 to 100 feet, or even larger, and reach a breaking pressure of 140 psi, well beyond the design limit of a ship.

The goal of the ongoing research is to understand how these waves are formed and then to predict their genesis. Mariners can be informed of areas to be avoided or be made aware of the likelihood of encountering a rogue while in particular conditions.

Thus far, no single explanation for the formation of rogues has emerged. Several theories appear reasonable, and most may be predictive enough to someday provide early warnings to mariners.

Here are the current rogue genesis theories. First, for some locations, *a mechanism of rogue wave formation is the focus of smaller waves by the shape of the coast or seabed. When conditions are right, the waves come together in phase.* This means the waves reach their maximums at the same time. Then their crest heights combine and cause a freak wave that might be double or more of the average wave height in the area.

Another mechanism for rogue formation occurs when *waves along a significant ocean current, like the Gulf Stream, are driven into an opposing current, or a strong wind from the opposite direction.* This produces a shortening of the wavelengths, or the distance between waves. As a result, the wave heights increase, and *oncoming wave trains compress together, potentially causing a rogue wave to form.* This mechanism accounts for the relatively common occurrence of rogues off the coast of South Africa, where a strong southwest ocean current is often opposed by the prevailing winds from the West.

This mechanism might have been the genesis of the rogue wave that capsized Philip Weld's trimaran. The Gulf Stream passes by Bermuda in the area where his accident occurred, and if there was an opposing wind at the time, the conditions could have been right for the formation of a rogue.

Yet another mechanism scientists have studied is *when a very large group of stable waves in cold water moves into warm water.* Because of the change in temperature, wave physics demands that the wavelength get shorter and the wave height get larger. The waves seek equilibrium by displacing energy from the front and back of the group toward the center. This action, then, can cause a rogue wave to form near the center of the wave group.

Physicists have been studying the behavior of waves for over a century. Several equations have been developed to describe how different wave types propagate over time. *One such model is known as the nonlinear Schrödinger (NLS) equation, created by Dr. Erwin Schrödinger.* A solution to this equation, discovered by British scientist Howell Peregrine in 1983, exhibits an unusual wave type that "sucks" energy from other waves, growing to a considerable size before *becoming unstable and collapsing.* This solution was tested in a wave tank by a team of scientists in 2011 at the University of Technology in Hamburg, Germany. A rogue wave consistent with the NLS was produced. So, this mechanism can also be a source of large, rogue waves in the open ocean.

As just discussed, the waters off Massachusetts are not immune to the occurrence of rogues. In 2011, two mariners from Germany were in their forty-five-foot sailing boat, Eva, southeast of Nantucket. They described the wave conditions at the time as high, but not rough, so they weren't concerned. Then a rogue wave hit their boat, capsizing it, breaking the mast, hatchways, and portholes, and tearing up part of the deck. The ship started to sink.

Fortunately for the crew, they were able to salvage their life raft along with a hand-held satellite phone and an old GPS unit. They contacted the Boston Coast Guard, which sent a helicopter to rescue them. Two hours later, the sailors were safe aboard and headed to a hospital on Cape Cod. *These mariners survived because they planned and had the right equipment.*

Now we know that rogue waves are real, large, unexpected, and dangerous. To make it worse, they can come from a direction other than the prevailing wind. Fortunately, rogues are relatively rare.

But when they hit a boat, there is little that can be done. Survival, as in the case of the German couple and Phil Weld, will depend on the preparations made before the voyage and a little luck. *Gloucester mariners, therefore, would do well to take appropriate precautions when they venture out to the open ocean.*

Update of the article

Additional incidents involving rogue waves - Rogue waves are substantial, unexpected, and often dangerous ocean surface waves. They are defined as waves whose height is more than twice the significant wave height (the average height of the highest one-third of waves in a given sea state). They appear suddenly and can be much larger and steeper than the surrounding waves.

In November 2020, a research buoy off the coast of British Columbia recorded what is considered the most extreme rogue wave ever measured relative to the surrounding sea state. It reached 58 feet in height, occurring in a sea with a significant wave height of only 20 feet, or about one-third of the size of the rogue wave. This highlights the unpredictable nature and immense size that these waves can achieve.

In December 2022, the cruise ship, Viking Polaris, was struck by a rogue wave while sailing in the Drake Passage[23] headed to Antarctica. Tragically, one passenger, a 62-year-old American woman, was killed when the wave broke through her cabin window. Four other passengers were injured. The ship had to cancel its scheduled Antarctic itinerary and return. This incident serves as a stark reminder of the danger that rogue waves pose, even to large vessels.

---

23 **The Drake Passage** is dangerous because of its unique geography, intense weather systems, powerful currents, and extreme isolation. It's the unavoidable gateway between South America and Antarctica, making it a rite of passage for sailors and scientists alike.

How likely is a Tsunami on Cape Ann?

First published in the Gloucester Daily Times on March 8, 2019

Tsunamis, Japanese for "harbor waves", can be devastating events. For example, on December 26, 2004, a series of tsunamis, reaching heights of up to 100 feet, struck the coasts of fourteen countries in the Indian Ocean, including Indonesia, Sri Lanka, India, and Thailand. *Astonishingly, almost 228 thousand people lost their lives.*

We live on Cape Ann, on the coast of the Atlantic Ocean, and could someday face the threat of a tsunami striking our shores. But just what is the likelihood of such an event? The short answer is that a moderate tsunami hitting Cape Ann is possible sometime in the next 100 years. We should, therefore, understand how tsunamis are created and where we should look for the one that might be coming at us.

Several events can generate a tsunami. They include massive underwater earthquakes, the eruption of underwater volcanoes, underwater landslides, and the least likely but the most devastating event, an impact into the ocean by a large asteroid or comet. For each of these events, the mechanisms generating tsunamis are similar. *The triggering event displaces massive amounts of water very quickly. The larger the amount of water displaced at the epicenter of the event, the larger the tsunami created.*

The 2004 tsunamis in the Indian Ocean were spawned by a major underwater earthquake of the *megathrust type*[24], about 100 miles off the west coast of Sumatra. Megathrusts are the most powerful earthquakes ever observed and occur when one *tectonic plate*[25] is forced under another. This mechanism accounts for the majority of deadly tsunamis. *In this*

---

[24] A *megathrust earthquake* is the most powerful type of earthquake, occurring at *subduction zones of tectonic plates.*

[25] A **tectonic plate** is a massive, irregularly shaped slab of solid rock, generally composed of both continental and oceanic lithosphere (Earth's outermost shell). The Earth's **lithosphere** is divided into several tectonic plates. These plates fit together like a jigsaw puzzle to form Earth's surface.

instance, it was a rupture along the fault between the Indian Plate and the Burma Plate.

During the Sumatra earthquake, a large area of the seabed had a sudden, vertical rise of up to twenty feet. This action displaced a massive amount of water, creating a huge underwater wave. In the deep water where the wave was created, *it appeared on the ocean's surface as a low, broad hump that was barely noticeable.* However, it began to spread away from the epicenter in all directions at speeds of up to 600 miles per hour. If it encountered a boat along its route, *the sailors would have barely noticed the wave since it was so broad and flat on the surface of the ocean.*

It became dangerous when it encountered the shallower water near the coastlines, interacted with the seabed, and slowed down to thirty or forty miles per hour. *This action caused the wave to build into the 100-foot monster that rushed ashore and went far inland* causing the major destruction and loss of life along all those hundreds of miles of Indian Ocean coastline.

On Cape Ann, we are far from the edges of any major tectonic plates where megathrust earthquakes are generated. Those quakes tend to be focused on the Ring of Fire of the Pacific Ocean. We are, however, *in a part of the Globe that is geologically susceptible to moderate earthquakes.* Such quakes could conceivably create a tsunami if they occur underwater, *but it would not be of the size that did so much damage in the Indian Ocean.*

The United States Geological Survey (USGS) is responsible for maintaining databases of earthquakes and using that information to understand the likelihood of such activity in the future. According to them, the Cape Ann area has had several minor quakes over the last 100 years. For example, between 1982 and 2003, there were five earthquakes generally in the same location, in the waters about 27 miles East of Rockport. Their magnitudes were low at only 2.2 to 3.6 on the Richter Scale, so no significant damage was done.

In the more distant past, however, a major earthquake off the coast of Cape Ann did indeed appear to generate a small tsunami. It happened on November 18, 1755. The upheaval was over 6.0 on the Richter Scale, so the quake was "strong" in size. In fact, it was about 1000 times more potent than the recent quakes off our coast.

The epicenter of the 1755 quake was estimated to be 24 miles east of Rockport, roughly in the same region as the more recent temblors. Due to the relatively small amount of development along the New England coast at that time, little damage was done, and no lives were lost.

There was no tsunami reported in our area from the 1755 earthquake, but it still may have created the first recorded tsunami in U.S. history. With about the right travel time to have come from Cape Ann, a small tsunami hit the northern islands of the Lesser Antilles chain, about 2000 miles south of us.

So, what is the probability that another significant earthquake will strike Cape Ann? The USGS forecasts that Essex County has a 19% chance of experiencing an earthquake of 5.0 magnitude or greater over the next 50 years. The probability of a 6.0 or greater, the size of the 1755 quake, drops to just 3.6%. Such an earthquake could cause considerable damage.

If the quake occurred under deep water in the general area of the 1755 quake, it could conceivably create a tsunami. But that event is much less likely than the occurrence of the earthquake itself.

Beyond underwater, major earthquakes and volcanic eruptions are also sources of tsunamis. Cumbre Vieja is a volcano on the island of La Palma in the Canary Islands. These islands are a Spanish possession in the Atlantic Ocean, located approximately 60 miles west of Morocco. No one knows when the next eruption of that volcano will occur. It could be soon, or it could be 100 years from now, but Cumbre Vieja will likely have a major eruption at some point in that time frame.

Scientists who have studied the possibilities indicate that the eruption will *likely cause the rocky, western flank of the volcano to be dislodged and slide into the ocean.*

Recent studies of Cumbre Vieja suggest that it is unlikely that the entire flank will collapse into the ocean at once. Instead, it will most probably slide into the water in stages over a significant period. Tsunamis could well be created as each section enters the ocean, but they will be much smaller than the feared mega-tsunami if the whole flank goes at once.

Scientists expect several smaller "warning" eruptions to occur before the big one that starts to dislodge the flank. And we will, therefore, be alerted well beforehand to the potential of a tsunami coming to us from Cumbre Vieja.

The tsunami-causing event with the lowest probability but the most significant impact in terms of wave size is an asteroid or comet striking the Earth in the Atlantic Ocean. The last enormous asteroid impact was about 66 million years ago.[26]

The probability of another asteroid of that size hitting us is exceptionally low. However, just in case, NASA has a mission to prepare for launching a vehicle that can destroy, or at least divert, such an asteroid, and plans are underway to have such a craft available.

Living on the coast of Cape Ann is one of life's joys that we all share. A downside is the potential for a significant-sized tsunami. The most likely candidate mechanism to cause a tsunami that would impact us is a major eruption of the Cumbre Viejo volcano. It could happen sometime in the next 100 years, but we should receive plenty of warning about its potential impact. So, sleep well.

---

[26] This was **the asteroid** that is widely believed to have caused the extinction of the non-avian dinosaurs. It struck Earth approximately 66 million years ago, creating the Chicxulub crater in the Yucatán Peninsula, Mexico. This date coincides very closely with the end of the Cretaceous period and the beginning of the Paleogene period.

Update of the article

Worldwide tsunamis since 2019 - The last five years have not seen a mega-tsunami on the scale of the 2004 Indian Ocean tsunami or the 2011 Tohoku tsunami. But there have been a few notable smaller tsunamis over this timeframe. Here are some noteworthy events:

July 29, 2021, Alaska, USA: A magnitude 8.2 earthquake near the Chignik area in Alaska generated a small tsunami that was observed on tide gauges in Alaska, Hawaii, and the U.S. West Coast. While it didn't cause widespread damage or fatalities, it was a significant seismic event that triggered a tsunami.

January 15, 2022, Tonga: The Hunga Tonga-Hunga Ha'apai underwater volcanic eruption caused a significant tsunami that was observed across the Pacific Ocean. Tsunami wave heights exceeding three feet were recorded in Tonga, Vanuatu, New Caledonia, Australia, the Cook Islands, New Zealand, Hawaii, Japan, Chile, Alaska, and California. *Tragically, there were deaths reported in Tonga and Peru due to this event*, and it caused substantial damage to nearby islands. Interestingly, the initial tsunami waves in some distant locations, like Japan, arrived earlier than expected, attributed to atmospheric pressure waves from the eruption pushing the water.

February 8, 2025 (Caribbean Sea): A magnitude 7.6 earthquake between Honduras and the Cayman Islands triggered a tsunami advisory for Puerto Rico and nearby islands. While sea-level fluctuations were observed (the largest being a 1.2-inch increase in Mexico), severe damage or widespread impact did not occur, and the advisory was later lifted. This event highlights that even large earthquakes in specific geological settings, such as the Caribbean, where fault lines don't always cause significant vertical displacement of seawater, don't necessarily lead to major tsunamis.

While the last five years haven't seen a mega-tsunami, these smaller-scale events serve as reminders of the ongoing risk and the importance of tsunami warning systems and preparedness.

Preparing for an Extinction Level Event

First published in the Gloucester Daily Times on January 20, 2017

A few weeks ago, NASA scientist Dr. Joseph Nuth warned that the world is currently unprepared for the potential impact of a large asteroid or comet colliding with Earth. The scientists refer to such objects as "impactors." Perhaps the ultimate catastrophe humans could ever face would be an extinction event caused by such a collision. It would result in a rapid and extensive decline in most of Earth's life forms.

The number of mass extinctions that have already occurred in history is larger than you might imagine. Scientists estimate that there have been at least five major mass extinctions over the last 440 million years, but depending on how the data are interpreted, there *could have been as many as twenty.*

*The effects of each major extinction event were that **65%** to 70% of all species then alive on Earth were lost.* It took millions of years after each disaster before new, evolved life once more flourished.

While the probability of such an event is low, the devastation is hard to contemplate. We can understand the level of destruction wrought by a large impactor by looking at what scientists have discovered about the results of an event that occurred in the past.

Most people have heard about the extinction event that happened 66 million years ago. The scientists call it the Cretaceous–Paleogene, or K-Pg, extinction event. It resulted in the doom of the dinosaurs, the reigning creatures at the time. Many scientists who have studied the evidence agree that *the cause was an impactor of six miles in diameter, about the size of the island part of Gloucester, east of the Annisquam River.* It struck in the waters off what is now the Yucatan Peninsula in Mexico. *The collision created an enormous crater that is 110 miles in diameter.*

The immediate effects of the impact of the Yucatan object are astonishing. The six-mile diameter asteroid or comet, travelling at *ten to twenty miles per second,* burrowed 12 miles into the Earth in less than a second. As a result, a vast, superheated cloud of dust and steam spread out from the area at supersonic speed, *equivalent to over one million hydrogen bombs going off simultaneously.*

An enormous amount of material, stones, and dust were ejected from the site. Some of the material was propelled out of the atmosphere and then returned as millions of blazing fireballs. Large wildfires were produced over major areas of the planet. In some localized areas, the raining, superheated material raised atmospheric temperatures to hundreds of degrees, incinerating all animal and plant life.

Another catastrophic result of the Yucatan impactor was the creation of a Mega tsunami. The usual tsunamis from earthquakes may produce a wave height of twenty to thirty feet when they land. The large Yucatan impactor, however, suddenly displaced an amount of water the size of a mountain. This displacement *created a tsunami thousands of feet in height as it hit the shore*! It may have been the largest tsunami in history. The rushing wave would have reached over one hundred miles into the land area that is now the southern United States, destroying all of the vegetation and animal life in its path.

In addition, the object's colossal impact created shockwaves in the Earth's crust that traveled large distances and resulted in numerous global earthquakes and volcanic eruptions. The local devastation caused by these events did not compare to the most far-reaching impact.

The longest-lasting and most destructive result of the event was the dust and particles that spewed into the atmosphere. Eventually, they came down and covered the entire Earth, creating a harsh environment for all life. Furthermore, *some of the dust remained in the atmosphere,* blocking sunlight for years. The result was the *rapid loss of all plant life that depended on photosynthesis.* This loss disrupted the dinosaurs' food chain, resulting in their extinction.

So, what can we do to protect ourselves from this ultimate disaster, and, for that matter, much more minor, say

Volkswagen-sized, asteroids that could destroy a city? Scientists at NASA are working on several related projects.

The first step is to identify and track the objects in our solar system that could come into a collision course with our planet. NASA's Jet Propulsion Laboratory is developing a space telescope called the Near Earth Object Camera (NEOC). *This spacecraft aims to identify asteroids on a potential collision course with Earth.* The goal would be to give scientists sufficient time to study the asteroid closely, allowing them to plan a mission to either destroy or divert it. If all goes well, NEOC will be launched in 2021.

Destroying an asteroid or a comet is not an easy task. To be effective, a nuclear explosion must be detonated in precisely the correct position, preferably in the impactor's interior.

Working with nuclear weapons experts, NASA scientists are contemplating an unmanned, two-part spacecraft that will do the job. *It is called the Hypervelocity Asteroid Intercept Vehicle.*

Here is how it would work. First, instruments aboard the first part, or leader craft, would *detect the target asteroid and select the best point of impact.* The leader craft would then separate from the follower craft and rush toward the asteroid's surface at high velocity. Just before it makes an impact, the leader craft would signal the follower craft to begin the detonation sequence of the nuclear explosive. *The leader craft would then smash into the asteroid with enough energy to create a deep crater.* The follower craft would enter the crater, and the nuclear device would detonate. *The powerful, subsurface explosion would shatter the asteroid into small fragments that would burn up harmlessly in Earth's atmosphere.*

Dr. Nuth, of NASA's Goddard Space Flight Center, said in his recent remarks that we must launch such programs now. If we ever spot a large asteroid on a collision course with Earth, we may not have enough time to stop it. He said it takes at least five years to prepare such a mission with a reliable spacecraft. We may get much less warning than that.

Remember that the probability of a mass extinction-level event is small, about one in 80 million. However, if scientists ever discover that a large object is on a collision course with Earth, it would be reassuring to know that NASA is prepared to prevent the disaster. To be ready, they need to follow Dr. Nuth's suggestion and build the asteroid-destroying spacecraft soon so that it can be launched with less than a year's notice.

Update of the article

The Near-Earth Object Surveyor—The NEO Surveyor (formerly known as the NEO Camera) is a planned space telescope designed by NASA to significantly enhance the detection and tracking of potentially hazardous asteroids and comets (near-Earth objects, or NEOs) that could pose a threat of impact on Earth. It was initially scheduled for launch in 2021 but is now expected to be launched much later.

The NEO Surveyor's primary mission is to discover and characterize a large number of NEOs, particularly those larger than 140 meters in diameter, which are big enough to cause significant regional damage in the event of an Earth impact. It aims to find at least two-thirds of these objects within its first five years of operation.

Unlike ground-based telescopes, which primarily rely on reflected sunlight, NEO Surveyor uses infrared sensors. This is crucial because many asteroids are dark and don't reflect much visible light, making them hard to spot. However, they emit heat, which is detectable in the infrared spectrum. This allows NEO Surveyor to detect objects regardless of their reflectivity, including those approaching from the direction of the Sun, which are difficult to observe from Earth.

The mission will gather sufficient data to accurately determine the orbits of the discovered NEOs, enabling scientists to predict their future paths and assess potential impact risks.

NEO Surveyor will also gather information about the physical properties of NEOs, such as their size, shape, and composition. This data is vital for understanding the nature of these objects and developing potential mitigation strategies if a threatening object is discovered.

By significantly increasing the catalog of known potentially hazardous near-Earth objects (NEOs), the mission will provide crucial data for NASA's Planetary Defense Coordination Office, helping to inform strategies for preventing future asteroid impacts.

The NEO Surveyor mission is a collaborative effort led by NASA's Jet Propulsion Laboratory (JPL), which is responsible for the overall development and management of the mission. Key partners and contractors involved in building the spacecraft and its instruments include:

BAE Systems: Providing the spacecraft for the NEO Surveyor mission, including the sunshade system, deployable aperture cover, and cryogenic thermomechanical components. They are also leading the observatory's assembly, integration, and testing.

Space Dynamics Laboratory (SDL) of Utah State University: Contributing critical components, including the advanced infrared imager.

Teledyne Imaging Sensors: Developing the infrared detector technology for the telescope.

University of Arizona: The principal investigator for the mission is based at this University.

Caltech/IPAC (Infrared Processing and Analysis Center): Responsible for processing and analyzing the scientific data collected by NEO Surveyor.

Laboratory for Atmospheric and Space Physics (LASP) at the University of Colorado Boulder: Supporting mission operations.

SpaceX: NASA has contracted SpaceX to launch the NEO Surveyor mission using a Falcon 9 rocket.

The current estimated launch date for the NEO Surveyor is no earlier than September 2027. This date has been adjusted over time due to several factors. The mission is slated to launch from Florida.

The Hypervelocity Asteroid Intercept Vehicle—**This concept system has been superseded** by the *Double Asteroid Redirection Test (DART)* mission, which successfully demonstrated kinetic impact as a method of asteroid deflection on the asteroid moonlet Dimorphos in September 2022. *It was a significant milestone in planetary defense. The following article covers this project.*

Protecting Earth from an asteroid impact

First published in the Gloucester Daily Times on February 27, 2025

If you have a good pair of binoculars, look closely at a full moon some summer night. You will see that it is pockmarked with craters. This is because it has been bombarded by asteroids and comets impacting its surface over billions of years. The Moon has no atmosphere and is geologically inactive; therefore, once a crater forms, it remains there unless another impact disturbs it.

Over the same period, the Earth has also been hit many times by these celestial bodies. Craters would also cover our planet, but our atmosphere, weather, and plate tectonics have eroded and reshaped the Earth's surface, so only a few of the most prominent craters have left traces of their existence.

One such impact occurred sixty-six million years ago. A massive asteroid about the size of the island portion of Cape Ann, traveling at about 50,000 miles per hour, struck in the ocean just off the Yucatan Peninsula of Mexico, creating an enormous crater 110 miles wide by twelve miles deep. The energy released was equivalent to millions of hydrogen bombs exploding simultaneously, causing planet-wide wildfires.

The massive tsunamis that were created were thousands of feet high, and, due to the resulting dust and debris in the atmosphere, a years-long "nuclear winter" ensued. The consequence was the extinction of about 75% of all life on Earth, including the dominant dinosaur populations.

Astronomers constantly monitor the skies, tracking any asteroids or comets that could potentially pose a threat to Earth. One such group is NASA's Asteroid Terrestrial-impact Last Alert System (ATLAS) in Chile. Late last December, *they detected a new asteroid in Earth's vicinity and dubbed it 2024 YR4 (YR4)*. Their observations and measurements found that it is about the size of Newell Stadium. When they calculated its orbit, they determined that there was a 1% chance it would impact Earth on December 22, 2032, just seven years away.

"An asteroid this size impacts Earth on average every few thousand years and could cause severe damage to a local region", said the European Space Agency in a space safety briefing. A strike by such a monster would have the energy release of a substantial hydrogen bomb and could destroy a city. The scientists are, therefore, paying close attention to this near-Earth visitor.

There has been only one such event in recent history. It happened on June 30, 1908, in a remote forest near the Tunguska River in Siberia, Russia. That asteroid was estimated to be about the same size as YR4. When it entered the atmosphere, it began to disintegrate, and a large explosion occurred at a five-mile altitude, releasing the energy equivalent of that of a fifty-megaton hydrogen bomb. *Eighty million trees over 830 square miles of uninhabited forest were felled.*

While the YR4 asteroid remained visible, scientists made astronomical observations and careful measurements. They used that information to fine-tune their calculations, and now they know with high certainty that YR4 poses no threat to Earth. It does have a 2% chance of impacting the Moon. *But what if it turned out that we did need to worry about an Earth impact?*

You may recall the 1998 movie "Deep Impact," which addressed this issue. In the film, the celestial object was much larger than YR4. It was about the size of the asteroid that killed off the dinosaurs and would be an Extinction Level Event (ELE) if it were to impact Earth. In the movie, the government attempted to address the problem by dispatching a spaceship equipped with a nuclear weapon to halt the oncoming monster. Blowing the asteroid to pieces did not work completely because the smaller pieces continued to come to Earth on the same trajectory and caused major damage.

In real life, NASA knows it is not "if" but "when" Earth will face such a problem, so they have been working for years to produce a practical solution. Their Planetary Defense Coordination Office sponsored the Double Asteroid Redirection Test (DART) experiment in October 2022. It was the first full-scale demonstration of asteroid deflection technology. Their spacecraft crashed at 14,000 miles per hour into the target asteroid, Dimorphos. This kinetic impact successfully altered the asteroid's orbit by more than the amount that would have been needed to divert it away from Earth. This marked humanity's first success in changing the motion of a celestial object in the desired way.

The DART kinetic impactor approach could have worked on YR4 if it were a problem. But if we were ever faced with a much larger celestial object, like the one that killed off the dinosaurs, we would need a "bigger boat," as the Police Chief said in the "Jaws" movie.

Sandia National Laboratories in New Mexico has shown how it can be done. They conducted a laboratory experiment to demonstrate how the radiation effects of a nuclear explosion strategically aimed at the object would produce enough force to deflect it from a fatal collision with Earth, without breaking it into dangerous pieces that would continue on the same trajectory.

In the face of a potential asteroid or comet impact, we have demonstrated our ability to innovate and prepare for the ultimate planetary defense challenge. Engineers have laid the groundwork for protecting Earth from catastrophic collisions through advanced space tracking systems, rigorous impact modeling, and the development of deflection technologies, such as kinetic impactors and nuclear radiation deflectors.

The presence of YR4 underscores the critical need for ongoing investment in space observation and mitigation strategies. By staying vigilant and proactive, we ensure that if an asteroid or comet ever sets its sights on our planet, we will be ready to avert disaster and safeguard the future of life on Earth.

Update of the article

NASA established the Planetary Defense Coordination Office (PDCO) in January 2016 - PDCO's ongoing activities encompass a wide range of projects aimed at identifying, tracking, and mitigating Near-Earth Objects (NEOs) that could potentially pose a threat to Earth.

The PDCO sponsors research and development of technologies and techniques for *deflecting or disrupting asteroids that could pose a threat to Earth.* This includes working with FEMA on emergency response planning for potential impacts and collaborating with the White House Office of Science and Technology Policy on national strategies for planetary defense.

The PDCO collaborates with *international partners* through the *International Asteroid Warning Network (IAWN*) and the *Space Missions Planning Advisory Group* (SMPAG), both of which are endorsed by the United Nations.

Essentially, the PDCO serves as the central hub within NASA for all matters related to planetary defense. It oversees projects and studies to find, track, and characterize Near-Earth Objects (NEOs) and develop methods to deflect or mitigate potential impact threats. *It sponsors a wide range of projects and programs, including:*

The Near-Earth Object (NEO) Observations Program – This program utilizes ground-based telescopes worldwide and space-based assets like NEOWISE[27] to search for, discover, and track NEOs, determining their orbits and physical characteristics.

Operating the Center for Near-Earth Object Studies (CNEOS) at JPL, CNEOS scientists calculate the orbits of known near-Earth objects (NEOs), predict their close approaches to Earth, and assess potential impact risks. If a NEO is found to pose a significant impact threat (greater than a 1% chance

[27] NEOWISE (Near-Earth Object Wide-field Infrared Survey Explorer) is an infrared, space-based telescope.

over the next 50 years), the PDCO is responsible for informing the government, the media, and the public. An example was the subject of this article, the YR4 asteroid.

Double Asteroid Redirection Test (DART) - This mission, a joint project between NASA and the Johns Hopkins Applied Physics Laboratory, successfully demonstrated the kinetic impactor technique by altering the orbit of the asteroid Dimorphos in 2022. *This was the first full-scale test of asteroid deflection technology.*

NEOWISE Project - The repurposed Wide-field Infrared Survey Explorer (WISE) telescope has been used to detect and characterize thousands of near-Earth objects (NEOs), particularly their sizes, using infrared measurements. *While the primary NEOWISE mission concluded in 2024, the data collected remains valuable.*

Bolide Data Release - The PDCO collaborated with the U.S. Space Force to release decades of data on bolide (bright meteor) events. This data helps scientists understand how asteroids break up in Earth's atmosphere and informs *impact modeling.*

Studies on Impact Mitigation Techniques—The PDCO has funded various studies exploring different methods of asteroid deflection and disruption, *including kinetic impactors, gravity tractors, and potentially (though with significant hurdles) nuclear options for short-warning-time threats.*

NEO Surveyor Mission (Launch no earlier than September 2027) - This is a crucial upcoming mission. This space-based infrared telescope is specifically designed to discover and characterize numerous potentially hazardous near-Earth objects (NEOs), including those that are currently undetectable with existing methods. *It aims to find at least two-thirds of NEOs larger than 140 meters within a decade of launch.*

Continued Research and Development of Mitigation Technologies- While DART was a significant step, research continues on various deflection and disruption techniques to prepare for diverse threats. This includes ongoing studies and potentially future demonstration missions (though none are currently formally announced beyond DART).

International Collaboration and Coordination - Maintaining and strengthening partnerships with international organizations and space agencies is a continuous priority for addressing global issues like planetary defense.

The PDCO's work is dynamic and adapts as new information about the NEO population is gathered and technological capabilities advance. *The focus is on a multifaceted approach* that includes detection, tracking, characterization, mitigation research, and international cooperation to protect Earth from potential asteroid and comet impacts effectively.

The science and vision behind a mission to Mars

First published in the Gloucester Daily Times on April 2, 2016

If you were over five years old on July 20th, 1969, you probably watched the live broadcast of the first Astronauts to land on the moon. And you have never forgotten the thrill. Hundreds of millions of people from around the world sat in rapt attention before their TVs, watching and listening to every word from the NASA feed. Since no one had ever done it before, not even NASA knew precisely what to expect. This made the attempt to land the unar Module, named the Eagle, even more dramatic. Some scientists speculated that the surface of the moon was covered with dust, many feet thick. The previous missions to the surface of the moon by the Surveyor spacecraft found that a manned landing could be made safely. But not all doubts were removed. If the dust were softer and deeper than planned, the Lunar Module's legs might sink into the surface. The craft would either have disappeared or tipped over. And that would have been the end of the mission and probably the end of the Astronauts.

We all held our breaths as the Lunar Module approached the surface. Fortunately, the dust layer was not a problem. Instead, we all watched as the Eagle landed safely on Tranquility Base. And, later that night, as Neil Armstrong set foot on the Moon's surface and said, "One small step for (a) man, one giant leap for mankind," it was an unforgettable moment in Earth's history.

Well, sometime in the 2030s, about twenty years from now, we will experience a similar event. Men will land on the surface of Mars; the first to ever set foot on another planet. It will be a much more difficult feat than landing on the moon. For one thing, the Moon is about 250 thousand miles from Earth. Whereas, Mars is about 250 million miles, or 1000 times further. But Charles Bolden, NASA Administrator, says that he is confident that we can accomplish the mission by that time. The various technologies necessary for a successful manned mission are all under development and are planned to be perfected in time for the mission launch.

Last year, the award-winning movie "The Martian" realistically highlighted these technologies. They include the ability to produce oxygen from the Martian atmosphere, to cultivate plants for food in a non-Earth environment, and to extract water from the Martian surface. Another is the development of deep space propulsion systems to transport astronauts to Mars and back to Earth.

The chemical rockets, with which we are all familiar, will not be adequate for the Mars mission. Yes, they will be used to get the spacecraft into Earth orbit. But, for the long, energy-consuming journey from Earth orbit to Mars, the limits of chemical rockets are exposed. The spacecraft, with the required payload and carrying the most energetic chemical propellant available, would still not have enough of that propellant to make the trip to Mars and back. Therefore, a new type of propulsion system is required to overcome this challenge.

In the movie "The Martian", an ion propulsion system was used to go to Mars from Earth orbit. This is precisely the type of system that NASA will use for the real-life mission. An ion engine uses electric power to ionize a gas, breaking down the gas atoms into positive and negative ions. Then, a strong electric field is employed to accelerate the positive particles (ions) out of the engine at super high speed. The result is a minuscule, but positive, thrust. In zero gravity space, however, even this tiny thrust can accelerate the spacecraft. The acceleration will be very slow. But it will be continuous, and over time, the craft can reach an extremely high speed. And that is the secret to the success of ion propulsion. It can operate continuously for months or even years with truly minimal fuel consumption compared to a chemical rocket.

The first NASA mission to use an ion engine was the space probe Deep Space 1, launched in 1998. A more recent application was the ion propulsion system for the Dawn spacecraft, launched in 2007. Dawn is now in orbit around the dwarf planet Ceres (diameter of 587 miles). Ceres is found in the asteroid belt between the orbits of Mars and Jupiter.

A key to utilizing ion engines is having sufficient electric power to operate them. To generate electric power, NASA has some options. The first option that comes to mind is solar panels. However, with today's technology, solar is unlikely to play a significant role in a Mars mission. This is because the further away from the sun the panels are, the energy available from the sun is reduced exponentially. Given that the Earth is 93 million miles from the sun, and Mars is about 250 million miles further out, the solar cells' production of electricity would only be a small fraction compared to the amount they generated in Earth orbit. To get the same amount of power in Mars orbit as in Earth orbit, the area of the solar panels would have to increase dramatically.

Solar cell technology development programs are working to increase the efficiency of the cells significantly. So, even with the smaller amount of solar energy available in Mars' orbit, more of that energy would be converted to electricity. If the efficiency improvements are significant enough over the next twenty years, solar panels may be able to play a more substantial role in a manned Mars mission.

Another power source used for many past Mars missions is a Radioisotope Thermoelectric Generator (RTG). It is a self-contained device, so it doesn't have the problem associated with solar panels. This generator uses a small amount of Plutonium-238, a relatively safe radioactive material. The isotope is naturally decaying and produces heat energy during the decay process. The heat is then transformed into electricity using a device called a Thermocouple. The amount of power produced by these units is adequate to operate the ion engine during the flight. Upon landing on Mars, the power sources can then be used to operate ground-based equipment.

It now appears, from a technological standpoint, that we can successfully undertake a manned mission to Mars. But should we? There is a perennial argument from people with a near-term view of the world that the resources required to go to Mars would be better applied to solving Earth's problems, like disease and poverty. For those with a long-term view of our future, however, there is no question. In their opinion, we need to leave Earth and colonize another planet; our long-term survival as a species may ultimately depend on it.

Update of the article

Elon Musk's plans for a manned mission to Mars - Elon Musk's gaze is fixed on Mars, not as a distant, romantic notion, but as a tangible destination, a second home for humanity. At the heart of this vision lies his SpaceX Starship, a colossal, gleaming vessel designed for full reusability, a key factor in making interplanetary travel economically feasible.

As of April 2025, SpaceX's Starship has conducted eight integrated flight tests. Out of these eight tests, four are generally considered to have achieved significant success, while the other four experienced various forms of failure during various stages of flight:

The successful tests achieved key milestones, including stage separation, booster return and landing (with spectacular tower catches), and controlled splashdowns of the ship. These flights provided crucial data for iterative development.

The failed tests experienced issues ranging from engine failures during ascent to in-flight explosions and loss of the ship before reaching the intended landing zone. Despite these outcomes, SpaceX considers these tests valuable for learning and making necessary improvements to the Starship design and operational procedures.

SpaceX's approach to Starship development involves rapid iteration and testing, with the understanding that failures are part of the learning process. Each flight, regardless of the outcome, provides valuable data that contributes to the ultimate goal of a reliable and reusable transportation system for Mars and beyond.

Unlike traditional space missions focused on fleeting exploration, Musk's ambition is *to establish a permanent, self-sustaining colony on Mars*, a safeguard against terrestrial existential threats. This isn't just about planting a flag; *it's about building a city.*

The path to this Martian city begins with *a series of uncrewed Starship missions*, a crucial phase for testing landing procedures and deploying the essential building blocks of a future civilization. These robotic pioneers will pave the way for human arrival, carrying with them the tools and equipment needed to *construct habitats, power sources, and life support systems.*

The cornerstone of Martian self-sufficiency is in-situ resource utilization, which involves extracting and processing local resources. Musk envisions vast networks of machinery, perhaps even augmented by humanoid robots, extracting water, ice, and atmospheric carbon dioxide to produce methane and oxygen, *the very fuel needed for return journeys and the foundation of a growing colony.*

This vision extends beyond mere survival. *The Martian colony is imagined as a thriving community*, a testament to human ingenuity and resilience. Infrastructure development is paramount, with *plans for robust power generation, comfortable habitats, and efficient resource processing facilities.* Musk's concept is not just about reaching Mars; it's about creating a new branch of human civilization, capable of thriving independently.

Musk has stated that SpaceX plans to launch the first *unmanned* Starship missions to Mars by the end of 2026. If these missions are successful in demonstrating the ability to reliably land Starships on Mars, then *manned* missions could begin as early as 2029, though Musk has indicated 2031 is more likely.

While ambitious, Musk's Mars plan is a dynamic and evolving project. SpaceX continues to push the boundaries of space technology, relentlessly testing and refining Starship. The challenges are immense, but the vision remains clear: to transform Mars from a distant, red dot into a vibrant, living world, a testament to humanity's enduring spirit of exploration and its determination to secure a future beyond Earth.

NASA's plans for a manned mission to Mars - NASA's overarching strategy for sending humans to Mars involves a multi-phased approach, building upon decades of robotic exploration and technological development. *A critical element is the Artemis program, which aims to establish a sustainable presence on the Moon first.*

This serves as a crucial steppingstone, allowing NASA to test and refine the technologies and systems necessary for longer and more complex missions to Mars. These include advanced life support systems, in-situ resource utilization (ISRU) techniques to harness resources like water and oxygen, and robust habitation modules.

The actual mission architecture for a human Mars expedition is still under development, but it will likely involve a complex series of launches and space rendezvous. NASA is developing the Space Launch System (SLS) rocket to carry the Orion spacecraft, which would transport astronauts to and from Mars. Deep space habitats are also being designed to provide a safe and comfortable environment for the crew during the long journey, which could take several months each way. Landing on Mars will require sophisticated entry, descent, and landing (EDL) systems to deliver astronauts to the surface safely.

NASA's current working timeline aims to send humans to Mars sometime in the 2030s. This ambitious goal requires continued progress in several key areas, including propulsion systems, radiation shielding, and closed-loop life support. International and commercial partnerships are also crucial to sharing the costs and expertise needed for such a complex undertaking. While the challenges are significant, NASA remains committed to its long-term vision of establishing a human presence on Mars, driven by the scientific potential and the desire to expand humanity's reach in the solar system.

Climate Change: science or religion?

First published in the Gloucester Daily Times on May 27, 2017

The study of how climate changes is a valid topic of science marked by a long history of scientific inquiry. It is clear from reading news accounts every day, however, that "Climate Change" has become the "religion" of many people, including some scientists. For these people, the historical scientific method for studying physical phenomena doesn't apply to Climate Change. This becomes obvious when you compare the characteristics of scientific inquiry to those of religions.

Science is based on a community of investigators conducting their research, sharing it with others, and working toward a consensus about the facts. A scientist refuses to accept results on authority. In fact, as Thomas Huxley, a famous British nineteenth-century biologist, said, "skepticism is the highest duty of a scientist, blind faith the one unpardonable sin."

Religion demands belief, science requires disbelief. We see that Climate Change demands belief. The believers say that the "science is settled" and anyone who dissents is a "denier", a derogatory term stolen from the description of the evil people who deny the Holocaust. So, not allowing dissent leads to blind faith, the "one unpardonable sin" in scientific study.

The renowned scientist and author Michael Crichton *was one of the first observers to identify environmentalism, in general, and Climate Change, in particular, as a religion.* In a meeting with the Commonwealth Club of San Francisco in 2003, he said, "Today, one of the most powerful religions in the Western World is environmentalism." "Increasingly, it seems facts aren't necessary because the tenets of environmentalism are all about belief. It's about whether you are going to be a sinner or saved. Whether you are going to be one of the people on the side of salvation, or on the side of doom. Whether you are going to be one of us, or one of them."

A good example of what happens to perceived Climate Change non-believers, or infidels, is Bret Stephens, a newly hired columnist for the New York Times (NYT). He was previously employed by The Wall Street Journal and won the Pulitzer Prize in 2013 for his commentary. He published his first column for the NYT last week, titled "Climate of Complete Certainty".

In his column, Stephens made several points that, on their face, would not be controversial in discussing a "normal", science-based topic. His primary theme was that it is absurd to blindly support Climate Change without listening to both sides of the argument. For example, he said that global "warming is indisputable". "as is the human influence on that warming." "Much else that passes as accepted fact is really a matter of probabilities." "That is especially true of the sophisticated but fallible models and simulations by which scientists attempt to peer into the climate future. To say this isn't to deny science. It is to acknowledge it honestly."

The strong reactions of some readers of The New York Times, all members of the Climate Change church, were unbelievable, to the point that their actions made national news for the next several days. Although not true, as confirmed in his comments above, in these readers' views, Stephens' column committed the mortal sin of "denying" Climate Change. Stephens received a massive amount of hate emails and tweets from Twitter Trolls. "When is the Times going to get rid of you?" was one of the milder examples. "The ideas ppl like @BretStephensNYT espouse are violently hateful & should not be given a platform by @NYTimes" was like many that espoused that Stephens should be shut down by the NY Times. Many of the other comments can't be quoted in a family newspaper.

Another common theme of these people was to order a subscription boycott of the NYT. "Each and every one of us should fully boycott the NY Times", "their actions are inexcusable. You cannot be an ostensible paper-of-record and allow a science denier to spread propaganda."

The most far-reaching reaction from readers was to create a petition on the Change.org site, asking the New York Times to fire Stephens. It had 25,000 signatures soon after the column was published and continued to grow.

Paul Krugman is a fellow NYT columnist who is a member of the Climate Change church. He has said in one of his columns, "may you be punished in the afterlife for being a denier. Denial is an almost unforgivable sin."

While Michael Crichton was among the first to point out the implications of the climate change debate, many other qualified observers joined him. Dr. Richard Lindzen of MIT said, "A surprisingly large number of people have concluded that all that is giving meaning to their lives is saving the planet by reducing their carbon footprint. People with no other source of meaning in their lives will defend their religion with jihadist zeal." He believes that the consequence is flawed science which then results in catastrophic public policy.

Even the former head of the EPA in the Obama administration, Gina McCarthy, agrees with the "religion" idea. Speaking to the Democrats in the California state Senate earlier this year, she said, "The challenge I think we have is for some reason climate change has become a religion, a politically induced religion instead of a scientific fact, that now we have to embrace and move forward on."

George Cardinal Pell is an Australian prelate of the Catholic Church, based in the Vatican. He also shares the concept of the Climate Change religion. He said, in disagreeing with Pope Francis' stand on Climate Change, "In the past, pagans sacrificed animals and even humans in vain attempts to placate capricious and cruel gods. Today they demand a reduction in CO_2 emissions."

Michael Crichton concluded his analysis with the following statement, with which I agree. "Because in the end, science offers us the only way out of politics. And if we allow science to become politicized, we will be lost. We will enter the Internet version of the dark ages, an era of shifting fears and wild prejudices, transmitted to people who don't know any better. That's not a good future for the human race. That's our past. *So, it's time to abandon the religion of environmentalism, and return to the science of environmentalism, and base our public policy decisions firmly on that.*"

Update of the article

Others who have stated that the Climate Change movement has the characteristics of a religion –

James Delingpole, a British journalist and writer, frequently uses the "Climate Change as a religion" analogy in his writings and commentary. Delingpole often portrays climate change activists as *zealots* and their beliefs as a form of fundamentalism. He emphasizes what he sees as *the moralizing tone of climate advocacy,* with its emphasis on guilt, sin (in the form of carbon emissions), and redemption through specific actions.

Patrick Moore, one of the founders of Greenpeace, is a vocal critic of mainstream Climate Change narratives. While not always using the term "religion," Moore has argued that *climate activism has taken on characteristics of a belief system,* with its prophets (scientists), sacred texts (IPCC reports), and moral imperatives. He contends that this *quasi-religious fervor can hinder rational debate and lead to policies that are not based on sound science.*

How ***Climate Change as a Religion*** *influences mainstream media coverage of weather events to create a sense of urgency in the population -* Al Gore authored a famous book and produced a documentary in 2006 on Climate Change, **An Inconvenient Truth.** In them, he presented scenarios of dramatic sea-level rise, suggesting the inundation of major coastal cities and substantial portions of landmasses. Some critics argued that these depictions were exaggerated and intended to create a sense of urgency. They were correct.

The extent of sea-level rise Gore depicted in his scenarios is dramatically higher than the 21st-century projections made by the United Nations Intergovernmental Panel on Climate Change (IPCC). For example, Gore spoke of a possible rise of 20 feet in sea level. *The IPCC says that sea-level rise is a gradual process that will occur over many decades and centuries. Their most likely sea rise over the 21st century will be only a fraction of Gore's twenty feet.* And most of that is not supposed to *occur until late in the century.*

Boston Harbor's tidal gauge provides an example of the *actual sea rise over the last fifty years.* The National Oceanic and Atmospheric Administration (NOAA) maintains a tidal gauge in Boston Harbor. Over the last fifty Years (1975–2025), according to NOAA's data, the relative sea level in Boston has risen at an average rate of approximately 2.97 millimeters per year since 1921. *This equates to about* ***5.8 inches*** *over the past 50 years.* Very different from Gore's twenty feet.

Here is an example of how the Mainstream Media (MSM) has been overplaying sea rise because of Climate Change to create a sense of urgency about the problem. A major storm, known as a bomb cyclone, hit Boston on January 4, 2018, resulting in a 15-foot tidal surge. *The news media reported the inundation as the effect of Climate Change. But was it?*

The total water level during the event was 15.16 feet above the mean low tide. However, that included an extremely high astronomical tide, as it was a "king tide" due to the Moon being in a very close orbit to Earth. Then, a storm surge was caused by the bomb cyclone's intense low pressure and strong onshore winds. Waves and local flooding dynamics were among the top factors.

5.8 inches of "sea rise", as recorded by NOAA, out of 15.16 feet of tidal impact, is only about 3% of the total water height. Therefore, *Climate Change was not a significant cause of that specific flooding event.*

It's true that, like concerns about exaggerated sea-level rise predictions, there are concerns about the mainstream media (MSM) overstating the connection between Climate Change and other specific weather events, such as hurricanes, droughts, and wildfires.

It's essential to recognize that the connection between climate change and individual weather events is a complex area of scientific research. While climate change is influencing weather patterns, it is not possible to isolate its precise contribution to a specific event. *Nuance can sometimes be lost in media reporting, which may focus on the most alarming aspects to capture public attention.*

Fact-checking a climate change zealot

First published in the Gloucester Daily Times on September 30, 2017

In the September 15th edition of the Gloucester Daily Times, an opinion piece was published titled *"Climate change another fact for team Trump to conveniently ignore",* by Ann McFeatters, a columnist for the Tribune News Service.

She says, "Scientists have figured out that hurricanes are stronger, bigger, and more dangerous because warming ocean temperatures spawn monster storms. Climate change, exacerbated by the burning of fossil fuels, is causing the oceans to warm. Storms will get worse."

McFeatters then says, "But, friends, this is not a matter of religious-like belief. Facts are facts or they wouldn't be facts." And she goes on to state that the Trump administration is ignoring these facts.

The problem is that her facts are not correct. When examining the federal government's historical hurricane data, the truth is that Harvey and Irma, two recent hurricanes, were not at all unusual. In fact, the official data available from *NOAA clearly shows that hurricanes are* ***not*** *becoming more numerous or more powerful when they make landfall in the U.S., negating the key point of her argument.*

NOAA's Hurricane Research Division maintains a database of all the Hurricanes that impacted the US mainland *from 1851 to 2016 (and now up to 2024).* Harvey and Irma have not yet been added to the official database, but news reports indicate that, at landfall, both had maximum winds of about 115 knots or *135 miles per hour.* This puts them at the *low end of Category 4* on the Saffir-Simpson Hurricane Wind Scale (131 to 150 mph*). Fourteen hurricanes in the NOAA Hurricane Database were stronger than these two, and ten others tied them.*

A more detailed inspection of the NOAA Hurricane database reveals that, over the 160-year period from 1850 to 2010, there was an average of 17 hurricanes per decade that made landfall in the U.S. Of these, an average of 5 per decade, or approximately *one-third, were Category 3 or higher.*

If one compares those long-term averages to the last forty years (1970 to 2010), there were fewer hurricanes, 15 per decade, compared to 17 per decade for the previous 160 years. And, on average, 5 per decade of category 3 or higher. This is the same average as that for the last 160 years.

The data, however, shows the opposite of what Ms. McFeatters and other climate change zealots claim. *Hurricanes are not becoming more frequent or of higher intensity due to climate change.*

Those who advocate that climate change is occurring at an alarming rate, is man-made, and that drastic, expensive actions need to be taken to avert disaster, are free to make their points. However, they must ensure that their "facts" are accurate, or they risk losing all credibility.

Update of the article

When examining the full 165-year record of Atlantic hurricanes maintained by NOAA (from 1851 to the present), *there is still no clear evidence of a significant long-term increase in either the frequency or intensity of hurricanes making landfall in the United States.*

While some datasets and charts from NOAA focus on more recent periods, from the 1950s onward, these can be misleading if taken in isolation, as they don't capture the full historical context. Some authors who want to support the climate change argument that hurricanes are becoming more frequent and intense use this "cherry-picking" of the data to support their claims.

The longer-term data reveal that the number of U.S. landfalling hurricanes, including major hurricanes (Category 3 or higher), ***has remained relatively stable,*** *with considerable natural variability over time.*

Periods of heightened activity, such as the 1880s, 1930s–1950s, and the early 2000s, have alternated with quieter decades like the 1900s–1920s and 1970s–1990s.

Multidecadal climate patterns, such as the Atlantic Multidecadal Oscillation, likely influence this cyclical behavior. Although early records are less complete, particularly for storms that remained at sea, hurricanes that made landfall were generally well-documented, lending confidence to long-term trends in landfalling storm data.

Some zealots predict that climate change will significantly influence hurricane characteristics, including increased rainfall, slower storm movement (which can increase damage), and possibly higher peak intensities. However, *the historical record through 2022 suggests that while certain aspects of storm behavior may be changing, the overall number of hurricanes striking the United States has not increased significantly.* Caution is warranted when interpreting short-term trends or cherry-picked datasets, and the full historical context is essential to draw accurate conclusions about long-term hurricane behavior.

How real is the Green New Deal?

First published in the Gloucester Daily Times on April 4, 2019

A few weeks ago, Senator Ed Markey (D-MA) and Congresswoman Alexandria Ocasio-Cortez (D-NY14) introduced a *Resolution* (not a bill) to Congress, H.Res.109 – "Recognizing the duty of the Federal Government to create a Green New Deal" (GND). The package focuses on creating a "green economy" that addresses climate change and simultaneously tackles *a range of social issues.* More specifically, it calls for a complete transition to clean, renewable energy, along with a host of far-reaching, socialist-inspired economic programs. All to be accomplished within ten years.

There are two primary reasons why Congress will never pass the GND. *First, the probable costs to implement such a program are astronomical.* Second, and just as problematic, the laws of physics inhibit its enactment. More about that later.

The GND document is a sprawling set of resolutions aimed at tackling climate change and promoting racial and economic equity. You may have heard that, if the GND were implemented, it would result in a ban on airplane travel in favor of high-speed rail, and the elimination of gasoline-powered cars in favor of electric-powered vehicles. Another implication would be the elimination of all cattle because they emit methane, a significant greenhouse gas.

The Resolution also includes several socialist-like goals, such as providing all Americans with housing, single-payer healthcare, and economic security. For example, according to Ocasio-Cortez, everyone would have a guaranteed minimum income paid by the government, even if they were "unwilling to work".

To be clear, this group of specific actions is not expressly included in the wording of the Resolution. They came, instead, from the documentation provided on Representative Ocasio-Cortez's website. She was trying to explain how the

Resolution's goals would be achieved in legislation. Her plan, however, received such a mocking reception that she soon withdrew it.

Instead, the GND Resolution outlines several national goals to be achieved over a ten-year mobilization period. In that sense, it would be akin to President Franklin Roosevelt's "New Deal" ideas, which he implemented in the 1930s to combat the Great Depression. This level of mobilization, the GND sponsors said, is necessary to prevent the potential "disaster" of Climate Change. *The top goal of the GND, therefore, is to reach net-zero greenhouse gas emissions in ten years.*

The cost of implementing the GND is hard to estimate at this point. Still, it is obvious that the costs would be enormous. One conservative think tank *estimated that it would cost $50 to $90 trillion over ten years.* Or approximately $5 to $9 trillion per year. *This compares to the 2020 total federal government budget of $ 4.7 trillion.* Larry Kudlow, the President's Economic Advisor, said, "The Green New Deal will literally destroy the economy."

There are many aspects of the GND resolutions that the Republicans find troublesome. One of the most problematic aspects is the goal to "meet 100% of the power demand in the United States through clean, renewable, and zero-emission energy sources" to achieve net-zero greenhouse gas emissions within the next ten years. This is where the laws of physics inhibit the implementation of the GND.

In the United States, we currently generate about 5% of our power from solar and wind sources. So, how could the GND supporters possibly have us make the switch to 100% renewables in such a short time?

Many of the people behind the GND believe that we are on the cusp of a technology-driven, energy revolution that will replace all hydrocarbon-based sources. They believe that solar and wind technologies are improving at a rate comparable to the historical rate of change achieved by information technology components and products. For example, the capabilities of computers and communication systems have grown exponentially over the last fifty years, resulting in dramatic performance increases while simultaneously experiencing significantly lower costs.

However, the laws of physics will not allow solar and wind to grow at similar rates in the future. For example, silicon photovoltaic cells, the basis of solar cells, have a physical boundary that results in a maximum conversion efficiency of only 34% for photons into electrons (*the Shockley-Queisser Limit*). We are currently at more than 26% in the best commercial products. Therefore, improvements are still possible, but nowhere near the level of exponential increase discussed by the GND proponents.

There is a similar physics boundary for wind turbines. The so-called *Betz limit* shows a maximum capture of only 60% of the kinetic energy from moving air. Commercial turbines today have a capture rate of more than 40%. Therefore, in the long term, exponential increases are not feasible in wind technology.

Another technology, not often discussed, but necessary to achieve 100% renewable power sources, is energy storage. When the sun doesn't shine and the wind doesn't blow, the solar and wind sources don't produce any power. *Today, when that happens, fossil fuel plants are utilized.* But, without the hydrocarbon plants, it will be *necessary to have some energy storage technology* to take over when required.

The primary issue is that none of the storage technologies currently being considered can match hydrocarbons in terms of energy density. Unlike hydrocarbons, renewable sources are captured dynamically from the sun and wind. Therefore, they cannot be stored and transported as easily as fossil fuels.

For example, one kilogram of crude oil contains almost fifty megajoules of chemical potential energy. In comparison, considering the physics of advanced lithium batteries, their maximum theoretical potential is only about 6 percent of crude oil! *This would imply that the size of such batteries to store utility amounts of energy would be prohibitively huge.*

There are several other types of energy storage technologies, such as those utilizing electric fields with capacitors and magnetic fields with superconductors, which have been considered for large-scale energy storage. However, currently, these technologies are only capable of energy storage of a mere 0.01 megajoules per kilogram, which is twenty times less than even the lithium battery, and a tiny fraction of the 50 megajoules per kilogram of hydrocarbons.

On March 26, 2019, Senate Majority Leader Mitch McConnell scheduled a vote on the Green New Deal resolution. Republicans opposed the Resolution, arguing that it proposed a "government takeover of the economy that could bankrupt the nation with an unrealistic goal of achieving net-zero carbon emissions within 10 years." Four Democrat Senators voted with the Republicans.

However, despite their vocal support for the Resolution among the Democratic Senators running for President in 2020, Harris, Booker, Warren, and Gillibrand did not vote in favor of the Resolution. Not even Senator Markey, who was the sponsor! Instead, they avoided taking a stand in favor of the GND and voted "Present".

To summarize, the Green New Deal would be enormously expensive and, due to the laws of physics, not even remotely practicable or possible over the next decade. *The Senate has unanimously voted it down, 0 to 57.* And, since it would probably receive a similar result in the House of Representatives, Speaker Pelosi is unlikely to schedule a vote in her chamber. Therefore, it is safe to say that "The Green Dream", as she once called it, is "dead".

Update of the article

Biden implementation of a Green New Deal - While the Green New Deal resolution was brought to the Senate floor for a procedural vote, it did not receive any affirmative votes. It did not advance further in the legislative process.

What happened after this vote is a good illustration of how politically and culturally divided the U.S. is in the 21st century.

During the Biden Administration (2021-2024), the principles embedded within the Green New Deal significantly influenced the legislative agenda. Many of the GND goals and proposals were incorporated into Biden's *Infrastructure Investment and Jobs Act of 2021 and the Inflation Reduction Act of 2022.* These landmark pieces of legislation *included substantial investments in areas aligned with the Green New Deal's aims, such as:*

- *Renewable Energy*: Tax credits, grants, and funding for solar, wind, and other clean energy technologies.
- *Energy Efficiency*: Programs to weatherize homes, improve building efficiency, and develop energy-efficient infrastructure.
- *Electric Vehicles*: Incentives for purchasing electric vehicles and investments in charging infrastructure.
- *Public Transportation*: Funding for expanding and modernizing public transit systems.
- *Clean Water Infrastructure*: Investments in replacing lead pipes and improving water quality.
- *Climate Resilience*: Funding for projects to help communities adapt to the impacts of climate change, such as extreme weather events.
- *Environmental Justice*: Initiatives that address pollution and environmental hazards in disadvantaged communities.

With the election of Donald Trump as President *however, many of these policies are being reversed and the funding redirected.*

Reversal of Biden's Climate Agenda:

Since returning to office in January 2025, President Donald Trump has made it a priority to dismantle what conservatives see as the overreach and inefficiency of the

Biden administration's climate agenda. Biden's Inflation Reduction Act, with its sweeping subsidies for electric vehicles and green energy companies, is being rolled back to reduce wasteful government spending and restore market-driven energy policy.

Legislation like the "One Big Beautiful Bill Act" aims to eliminate costly tax breaks for EVs and solar energy, which critics argue unfairly distort the energy market and favor elites.

In its place, the Trump administration is championing energy independence through expanded oil, gas, and coal production, seen by many conservatives as essential to national security and economic stability. By withdrawing again from the Paris Agreement, Trump is emphasizing American sovereignty over what he and his supporters view as ineffective global climate accords.

These efforts are not without controversy, but many on the right see them as a long-overdue correction to radical environmentalism. The administration argues that Biden-era climate regulations have burdened small businesses, increased energy costs for American families, and outsourced critical manufacturing to foreign nations like China. Reversing these trends is expected to strengthen the domestic industry, protect jobs in traditional energy sectors, and prioritize reliable power sources over unproven technologies.

While critics warn of environmental fallout, conservatives maintain that responsible energy development and innovation can coexist without sacrificing economic freedom or imposing top-down mandates. Legal pushback and court rulings may slow implementation, but for many on the right, Trump's policies represent a reaffirmation of America-first principles and a rejection of globalist climate orthodoxy.

Our Energy Future

First published in the Gloucester Daily Times on October 15, 2015

Our current primary sources of energy are fossil fuels and will remain so for many years to come. However, the supply and production of fossil fuels, including oil, coal, and natural gas, are going to eventually peak and then start to decline. Given the growing energy demands of the world, the cost of these fossil fuels will, therefore, increase dramatically with the decrease in supply. Some analysts call this event "peak oil". Depending on the evolving discovery and recovery technologies, it may be fifty or a hundred years or more, but it will come. Then what?

History shows that entrepreneurial spirit, technological readiness, and economics will converge on a new power-generating system that meets the market's needs at the most competitive price. For example, it was *Thomas Edison who, in 1882, combined the technologies of electric power generation and the light bulb to create illumination for homes and businesses, achieving superior results at a competitive cost compared to gas.* Therefore, a new power-generating system will be part of our energy future.

Current alternative energy sources, such as solar panels mounted on the ground or rooftops, or wind farms composed of giant wind turbines, are limited by the environmental conditions they require to produce power. *When it is dark or cloudy or the wind isn't blowing, the solar panels and the wind turbines are lifeless.* With these technologies, therefore, there *will always be the necessity to have enough generating capacity from other, steadier sources of energy production to meet our total needs at any point in time.* So, these part-time technologies are not the long-term answer.

Nuclear power is steady and does not produce harmful emissions; however, it has several negative aspects that make it an unlikely *long-term solution.* For example, although it is improbable with modern technology, disastrous accidents are possible, such as those that occurred at Three Mile Island, Chernobyl, and Fukushima. Furthermore, nuclear power plants produce radioactive waste that must be stored safely for extended periods.

What, then, might the new, alternative energy sources be, and how and when will they be available? No one can say for certain, but several possibilities on the horizon could eventually meet all the conditions for steady, safe, efficient, and economical energy generation. *I will focus on just two of the more interesting ones, Fusion Power (Fusion) and Solar Power Satellites (SPS).*

Fusion power is generated by creating super-hot plasma, a highly excited state of a gas, where energy is released from the fusion of the nuclei of billions of deuterium atoms. This energy is converted into heat, which is used to create steam that runs an electric generator. The advantages of Fusion are that it produces no atmospheric emissions, the fuel is cheap and abundant, and there is no chance of a catastrophic accident. The amount of radioactive waste is minimal. So why don't we have it already?

The primary problem has been in containing the hot plasma long enough to produce more energy output than was invested in creating it. Scientists have been working on an approach for decades. While progress in solving the technological problems has been slow, it has been steady. Scientists in many countries are now involved in the work. However, it will likely take a few additional decades before an economically viable commercial version of a Fusion power reactor is available.

The latest and largest project designed to achieve the goal of significant net energy from a Fusion system is the ITER (International Thermonuclear Experimental Reactor) being constructed in the South of France. The United States, along with the E.U. and five other countries, is sponsoring the project. This year, 2015, the test device, called Tokomak, will start to be assembled. It is expected to be completed in 2019, with the first tests using super-heated plasma scheduled for 2020.

The primary goal of the tests is to *deliver ten times the amount of power that the reactor consumes as output.* And then to scale the technology up to produce a demonstration Fusion Power plant for commercial use. The construction of the commercial plant is planned to start in 2024.[28] So, the last half of the twenty-first century could see the availability of cheap, abundant, and safe energy from this new source. For more information, see www.iter.org.

A second possible energy source to replace fossil fuels is Solar Power Satellites (SPS). With this alternative, large satellites are placed in geosynchronous orbit over the Earth, *featuring up to six square miles of solar energy-collecting panels.* Each satellite of this size would produce the same amount of power as two land-based nuclear plants. The collected solar energy is converted into another type of energy, such as microwave or infrared laser, and wirelessly beamed to collectors on the ground, where it is converted into electricity and distributed to users.

The major benefits of this system would be that solar energy is free and abundant, its collection in space would be uninhibited by atmospheric conditions or the day-night cycle, and there would be minimal environmental issues. Another benefit would be *that the energy could be beamed to places where it is needed,* either to existing distribution infrastructure, or to battlefield locations for the military, or to remote locations like the Arctic and Antarctic regions.

*The concept for SPS was conceived and patented by Dr. Peter Glaser of Arthur D. Little, Inc. in the **1960s**.* The International Academy of Astronautics (IAA) recently published the results of a three-year study that assessed the role that SPS might play over the coming decades (www.iaaweb.org). A major conclusion was that such systems are indeed technically feasible. Some technologies still need further development, but no breakthroughs are required. The central question is economic viability.

Significant cost reductions would need to be achieved in several technologies. For example, the cost of Earth-to-orbit

---

[28] This article was written in 2015. The ITER staff was overly optimistic. See, the next article on the current status of fusion power.

transportation is crucial, including transporting large payloads to geosynchronous orbit, approximately 22,000 miles above the Earth.

Several commercial firms are exploring such systems, and the IAA believes that they are likely to evolve over the next few decades to become cost-efficient.

Given the size of the solar cell arrays that will be required, a significant decrease in the cost of solar cell production and/or a substantial increase in their efficiency would also be crucial to making the systems economically viable.

The Japanese are currently at the forefront of exploring this promising technology. For example, the Japan Aerospace Exploration Agency (JAXA) is planning a series of ground and orbital demonstrations of the various technologies needed to produce an SPS. They expect to have a commercial system in place by the 2030s that will produce approximately one gigawatt of power, comparable to the size of a typical nuclear power plant.

In conclusion, as the economics of traditional energy production changes and other technologies evolve, commercial companies will bring new sources of power online to meet the world's needs. Both Fusion and SPS are good, long-term possibilities.

Update of the article

Progress of fusion power research - The goals for the ITER fusion power project, as listed in this 2015 article, were to be achieved by now, but they have still not been accomplished. This project has been slow throughout its history and has consistently pushed its objectives beyond the scheduled time.

The following article on page 254, "Finally, the dream of fusion power!", contains more recent coverage of fusion research and updates the status of ITER.

Progress on Space-Based Solar Power - The timeline for operational Satellite Power Systems is challenging due to numerous technological and financial hurdles that must be overcome. However, some governments and private companies have stated their goals and made projections.

Near-term Demonstrations (within the next 5-10 years): Several entities aim for smaller-scale demonstrations in this timeframe. Japan Aerospace Exploration Agency (JAXA) has a small demonstrator launch planned for 2025. Space Solar, a UK-based company, aims for a megawatt-scale commercial system within five years (by 2030). China plans to launch small to medium-sized test satellites by 2025 and aims for a megawatt-level system by 2035.

Pilot and Prototype Systems (within the next 10-20 years): Following successful demonstrations, the next phase would involve larger pilot plants. European Space Agency's SOLARIS program envisions a pilot plant by 2035. Space Solar, a UK-based company, aims to achieve gigawatt-scale systems by 2037. China seeks to develop a 200-ton Space-Based Solar Power (SBSP) station capable of generating megawatts by 2035 and a commercially viable station by 2050.

Commercial-Scale Deployment (beyond 20 years): Large-scale, commercially viable Space-Based Solar Power systems capable of significantly contributing to terrestrial energy grids are likely to be further out, potentially beyond 2040. The UK's Space Energy Initiative suggests an operational system could be developed by 2040.

Long-Term Outlook for SBSP Technology: The long-term outlook for SBSP technology is promising, driven by several factors:

> *Continuous and Abundant Energy Source*: SBSP offers the potential for a continuous, 24/7 energy supply, unaffected by weather conditions or Earth's day-night cycle. The amount of solar energy available in space is significantly greater than on Earth.
>
> *Clean Energy Transition*: As the world seeks to decarbonize and transition to clean energy sources,

SBSP could become a crucial component, providing a baseload renewable energy supply.

Technological Advancements: Ongoing advancements in solar panel efficiency, wireless power transmission, space launch capabilities (especially reusable rockets, such as those developed by SpaceX), and in-space assembly techniques are making Space-Based Solar Power (SBSP) increasingly feasible.

Energy Security and Independence: For nations seeking energy independence and a secure, reliable energy source, SBSP offers a potential solution.

Space Economy Development: The development and deployment of SBSP infrastructure would drive innovation and growth in the broader space economy, including manufacturing, launch services, and robotics.

However, significant challenges remain:

High Initial Costs: The upfront investment required for developing, launching, and assembling large-scale space-based solar power systems *is substantial.*

Technological Hurdles: Efficient and safe wireless power transmission over long distances, the assembly of massive structures in space, and ensuring the longevity and reliability of space-based hardware in a harsh environment are complex technical challenges.

Environmental Concerns: The potential impacts of large-scale microwave or laser power transmission on the environment and human health require careful assessment and mitigation. Space debris is also a growing concern for extensive space infrastructure.

Regulatory Framework: International agreements and regulatory frameworks will be necessary for the deployment and operation of SBSP systems, including the allocation of spectrum for power transmission.

Despite these challenges, the increasing global focus on clean energy and the rapid advancements in space technology suggest that Space-Based Solar Power has the *potential to become a significant energy source in the long term*, possibly contributing substantially to global energy needs in the latter half of the 21st century and beyond. My former colleague, Dr. Peter Glaser, inventor of the Solar Power Satellite system, would have been ecstatic with the progress being made.

Finally, the dream of fusion power!

First published in the Gloucester Daily Times on March 31, 2022

"Climate Change" has evolved from a scientific, environmental issue into a political football. On one side of the argument, some congressional groups advocate for overhauling the economy and spending trillions of dollars on their Green New Deal, a combination of renewable energy projects and a broad range of far-reaching social spending programs. While others in Congress don't believe the problem is severe enough to warrant drastic economic upheaval.

There is, however, one solution that both sides can agree upon. The achievement of a working, practicable, fusion reactor. It is the ultimate answer to Climate Change as well as our long-term energy needs.

Such a power source does not use fossil fuels and produces no "greenhouse gases". The primary fuel used in the reactor comes from sea water and is cheap and abundant. The reactor produces minimal radiation and does not have the spent fuel disposal problems associated with nuclear power. And, unlike nuclear power, it can never "run away" nor explode.

It produces far more energy in a much smaller footprint than solar or wind sources and is always "on", independent of time of day or weather conditions. In other words, it is the perfect, inexpensive, pollution free, sustainable, and safe power source.

Although scientists and engineers have known all of this for the last seventy years, the goal of producing a commercially viable fusion reactor has been highly elusive. *Now, however, scientists at the Massachusetts Institute of Technology, along with a commercial spinoff company, Commonwealth Fusion Systems, believe that they are very close to solving the myriad of problems and producing a workable solution.*

I first became aware of the possibility of fusion power back in the early 1960s. I was working in my first position as a physicist and, as part of my job, I had to become familiar with plasma physics (a super-hot gas of ionized particles). MIT was very active in the field, and I took advantage of their library and lecture series resources.

At the time, researchers at the Institute working on fusion power predicted that it would take about thirty years to produce a commercial power plant. The running joke in the community since then has been that the "thirty-year" prediction remained constant, each year for the next sixty years. That is, until now.

The world has set its hopes for a successful fusion reactor on a multinational project named the *International Thermonuclear Experimental Reactor* (ITER). This development is sponsored by thirty-five nations, including the United States, European Union countries, China, and Russia. The experiment site is in southern France. The fusion device they are building is called a *Tokomak* and is based on a highly refined version of a Russian reactor design from the 1950s. Essentially, the Tokomak creates and retains a high-density, super-hot plasma, or ionized gas, in a toroidal (donut-shaped), magnetic "bottle".

The plasma is so hot that there is enough energy to fuse hydrogen atoms. This means that the nuclei of two hydrogen atoms combine to form one atom of helium. This is the reaction that drives the production of power from the Sun. Accompanying this reaction is the release of highly energetic particles called neutrons. *These neutrons can then be captured and used to heat water, producing steam that ultimately powers ordinary generating equipment to produce electrical power.*

The ITER project began in 1988 and has made slow, albeit expensive, progress. Now, thirty-four years since the effort was initiated, only about 80% is complete. It is the most expensive science experiment of all time and is expected to cost between $45 and $65 billion by the time it is finished.

The ITER structure is many times larger than any other Tokamak ever built. The colossal size is necessary, given the capabilities of current magnet technology. The magnets must be giant to create a strong enough magnetic field to contain the self-sustaining, stable plasma that produces more energy than is needed to maintain it. This is the primary goal for demonstrating a workable design. *That achievement for ITER is planned, realistically, for 2033, eleven years from now.*

On the other hand, the MIT team believes they can reach these goals sooner and at significantly lower cost with its own, much smaller Tokomak named SPARC. *The key breakthrough for SPARC is its use of a new, superconducting magnetic material, developed at MIT, which can produce much higher magnetic fields per unit volume.*

The MIT magnets produce 54% higher field strength than the ITER magnets, and yet are significantly smaller and much less expensive. For example, the MIT magnet design has produced a field strength of 20 Tesla (a scientific measure of magnetic field strength). This compares to the ITER magnets that only create 13 Tesla.

It is this achievement that enables the entire MIT Tokomak to be built significantly smaller, less costly, and more effectively than the massive ITER device.

SPARC is going to be about one fortieth the size of ITER. It began construction a few months ago in December 2021. It is located close to Cambridge, at Devens, the site of the former U.S. Army Fort in Harvard, Massachusetts.

The scientists expect SPARC to be the first device in the world to demonstrate successful fusion power. The planned date is in 2025, eight years before the goal is planned to be achieved for ITER. It is worth noting that ITER has already adjusted this goal multiple times and is likely to do so again.

The SPARC experiment is expected to generate 50 to 100 Megawatts of fusion power, or approximately 10 times more than the power required to create the plasma. Then, the experience with operating SPARC for a year or two will provide the scientists with the guidelines they need to design and produce a full-sized fusion power plant with a capacity of 250 to 1000 megawatts of electricity. Such a plant will likely have a smaller footprint but produce more power than today's average fossil fuel power plant.

By the time the first fusion power plant from MIT-Commonwealth Fusion Systems is in operation, it will be clear to Congress, the Climate Change lobby, and the rest of the energy industry that fusion power is a reality. It will be crucial to implement the change as quickly as possible with an Apollo-type program. *Everyone will be delighted to go along.*

Update of the article

Outlook for the ITER project - As of April 2025, the ITER fusion project in Cadarache, France, has encountered significant delays, extending its operational timeline into the future.

Initially, ITER aimed to achieve "first plasma", the initial test of its fusion system, in December 2025. However, due to technical challenges, including defects in key components and the need for extensive repairs, the project has revised its schedule. *The new plan sets the target for first plasma to occur no earlier than 2033, with full plasma current expected in 2034.*

Subsequent operations using deuterium-deuterium fuel are scheduled for 2035, and deuterium-tritium fusion experiments, the stage at which ITER aims to demonstrate a net energy gain, are expected to begin around 2037, marking a critical milestone in the project's mission to achieve sustained nuclear fusion and pave the way for future commercial fusion reactors.

The major, very disappointing delays in the program stem from several factors:

Technical Issues: Defects were discovered in critical components, necessitating disassembly and repairs.

COVID-19 Pandemic: The global pandemic disrupted supply chains and on-site work, contributing to schedule setbacks.

Complexity of the Project: ITER's unprecedented scale and complexity have led to unforeseen challenges in construction and assembly.

These issues have also led to a significant increase in the project's budget, with costs rising by approximately $5 billion, bringing the total estimated cost to around $25 billion.

Despite these setbacks, ITER aims to demonstrate the feasibility of fusion as a large-scale and carbon-free source of energy. While the revised timeline represents a substantial delay, the project's progress continues, with ongoing efforts to address technical challenges and advance towards operational milestones.

However, other fusion power programs, such as the MIT-Commonwealth SPARC project, could demonstrate fusion power years before ITER.

Status and Outlook for the MIT Commonwealth Fusion Systems project -

As of April 2025, MIT- Commonwealth Fusion Systems (CFS) is making significant strides in its mission to commercialize fusion energy through its SPARC tokamak project in Devens, Massachusetts.

Current Status of SPARC

- *Assembly Underway*: CFS has commenced the assembly of the SPARC tokamak, marking a pivotal phase in the project's development. This follows the successful construction of their Devens campus, which includes a 165,000-square-foot manufacturing facility dedicated to producing high-temperature superconducting magnets essential for SPARC's operation.
- *First Plasma Timeline*: The company anticipates achieving "first plasma", the initial activation of the fusion device, in 2026. This milestone is a critical step toward demonstrating net energy gain, where the

energy output from fusion exceeds the energy input required to sustain the reaction.

- *Funding and Support*: CFS has secured over $2 billion in funding from prominent investors, including Breakthrough Energy Ventures, Tiger Global, and Google. This financial backing underscores confidence in CFS's approach to achieving practical fusion energy.

Outlook and Future Plans

- *ARC Power Plant*: Building on SPARC's anticipated success, CFS plans to develop ARC, a grid-scale fusion power plant projected to generate 400 megawatts of electricity. The ARC facility is slated for construction in Virginia, with operations expected to commence in the early 2030s.
- *Regulatory Landscape*: The U.S. Nuclear Regulatory Commission has established a distinct regulatory framework for *fusion* energy, separate from that of traditional nuclear *fission*. This differentiation is expected to streamline the development and deployment of fusion technologies.
- *Strategic Importance*: Fusion energy is increasingly viewed as a vital component of the global transition to clean energy. U.S. officials, including Virginia Governor Glenn Youngkin, have emphasized the need to accelerate fusion development to maintain energy leadership and meet the increasing electricity demand.

In summary, Commonwealth Fusion Systems is progressing steadily toward demonstrating net-positive fusion energy with SPARC, to achieve the first plasma in 2026. The subsequent ARC power plant represents a significant step toward commercial fusion energy, with plans to supply substantial power to the grid by the early 2030s. *While challenges remain, CFS's advancements position it as a frontrunner in the pursuit of practical fusion energy solutions.*

Conclusion to Part IV

As we bring this section of *Our Dynamic Planet* to a close, one message resonates across all the chapters: our planet is not a static sphere but a complex, ever-evolving system, constantly shaped by natural forces and human ambition. From the shifting lines of Earth's magnetic field to the massive energy lurking in rogue waves and undersea earthquakes, the physical world around us is filled with dynamic processes that can both inspire awe and pose profound threats. The likelihood of a tsunami off Cape Ann, for instance, forces us to consider how local geology and global tectonics intersect in shaping coastal risks. At the same time, the invisible dance of the magnetic poles reminds us that Earth's inner workings are far from settled.

But Earth's story is no longer limited to what lies beneath our feet or across our oceans. Humanity's gaze has turned skyward, driven by the same instinct to explore and understand. Our examination of Mars missions highlights how Earth's changing conditions are motivating technological visions that reach beyond this world, while discussions around asteroid deflection and extinction-level preparedness reveal a sobering truth: our survival depends as much on anticipation as on reaction. We are not simply passengers on this dynamic planet; we are also stewards and, increasingly, defenders against cosmic hazards.

In parallel, the challenge of Climate Change brings an entirely different kind of dynamism, one rooted not in cataclysmic events but in slow, systemic shifts that are harder to perceive yet far more enduring. As the climate debate becomes entangled with ideology, the science remains clear: understanding and mitigating human impact on Earth's systems is essential. Solutions like fusion energy, long considered the realm of science fiction, are beginning to materialize, offering hope for a sustainable future. The chapters on fusion and our broader energy future offer a glimpse of how innovation and science might counterbalance the environmental disruptions we face.

Ultimately, this collection of chapters illustrates that to live responsibly on this dynamic planet, we must embrace a dual perspective: one eye on the deep, often invisible forces of nature, and the other on our capacity for innovation and foresight. The Earth is a planet in motion, both geologically and atmospherically, and our response to its rhythms will define the future not just of civilization, but of life itself. What lies ahead may be uncertain, but with science as our guide and vigilance as our duty, we can prepare for both the challenges and the possibilities of a changing world.

Part V

Advanced Inquiry: Medicine, Metaphysics, and Modern Physics

V. Advanced Inquiry: Medicine, Metaphysics, and Modern Physics

Introduction to Part V

In this section, we enter a domain where the boundaries of science blur and deepen, where medicine, consciousness, spirituality, and the most perplexing discoveries of modern physics begin to converge. A common impulse unites these chapters: to press against the limits of current understanding and ask questions that once belonged solely to religion or philosophy. From the cellular level to the cosmos, from life's final breath to the quantum realm, this section invites readers to consider a universe far more mysterious, and perhaps more meaningful, than traditional science has allowed.

We begin with provocative explorations of faith-inflected science. What does empirical research say about ancient relics like the Shroud of Turin? Is it possible to detect the signature of a designer in the fabric of nature? These chapters don't preach conclusions but rather examine the evidence and arguments with intellectual rigor. Whether one is skeptical or devout, these inquiries demonstrate that some of our oldest questions—about the origin and purpose of life are now being reexamined with the tools of modern science.

Turning inward, the section then examines the remarkable advancements in medical science and the biology of consciousness. Regenerative medicine is poised to transform the way we heal, and breakthroughs in longevity research hint at the radical possibility of extending human life far beyond its current limits. Alongside these advances, we enter the deeply personal space of near-death experiences, brain imaging, and the dream state, phenomena that challenge a purely mechanistic view of the mind and invite us to reconsider the nature of consciousness itself.

Finally, we move from the inner world to the outer reaches of reality, where physics reveals a universe that is stranger than any myth. Time travel, once the stuff of science fiction, now finds theoretical grounding in modern physics. Quantum mechanics continues to confound our understanding of space, time, and causality. In these final chapters, we encounter a cosmos that resists simplification, a reality that might be entangled with consciousness itself. Together, these essays illuminate a profound truth: to understand the universe, we must explore not just matter but meaning.

The scientific evidence for the Shroud of Turin

First published in the Gloucester Daily Times on March 28, 2024

Since 1578, the religious artifact known as *The Shroud of Turin* has been preserved in the Cathedral of San Giovanni Battista's royal chapel in Turin, Italy. The ancient linen cloth is quite large, measuring approximately fourteen feet long by four feet wide. It has a full-length, realistic image of a crucified man, both his front and back. The man displays wounds exactly as described in the Gospels for the crucifixion of Jesus of Nazareth.

"Is it a forgery or is it genuine?" is the question that many have asked over the centuries. The scientific investigation of the Shroud began on May 28, 1898, when an Italian lawyer and amateur photographer named Secondo Pia was granted permission to take the first photographs of it. Upon developing the images, Pia was astonished to discover that the figure on the cloth appeared as a photographic negative. In other words, the Shroud itself acts like a photographic negative, where light and dark areas are reversed. As a result, *when Pia developed his negatives, they produced a positive image*, revealing the figure on the Shroud in striking detail.

With this finding, it became increasingly difficult to explain away the Shroud as a forgery, as it was hard to imagine how a medieval counterfeiter could have created such an effect.

Pia's photographs created a sensation, sparking renewed interest in the authenticity of the Shroud. But it took eighty years, until 1978, for that fascination to reach its peak. It was then that the official investigation, known as The Shroud of Turin Research Project (STURP), convened an international team of approximately forty scientists and technicians in Turin to examine the Shroud. The team leader was John P. Jackson, a physicist from the U.S. Air Force Academy in Colorado Springs, Colorado.

The team identified several key findings. First, they determined that *the image was embedded in the fibers of the cloth.* It is not a surface painting or drawing as would be expected for a forgery. Second, the *wounds shown on the man in the Shroud are consistent with death by crucifixion.* This includes nails through the wrists, *not the hands.* The Romans knew that a nail in the hand could not bear the weight of the body hanging on the cross. Third, *the blood on the Shroud is real human blood of the rare AB type.*[29]

The fourth key finding was the most astonishing. The total image of the Shroud contains all the necessary information to display the body in three dimensions. It shows the 3D contours of the body underneath the cloth, including the nose, mouth, and cheekbones, as well as the positions of the arms and legs that are consistent with a human body in rigor mortis.

Since those tests in 1978, the three-dimensionality of the image has been confirmed by other, even more sophisticated scientific methods. This characteristic of the image is one of the most convincing arguments for its authenticity, as it is challenging indeed to explain how a medieval forger could have created it.

Considering the physical effects on the cloth that produce the image, a proposal by some scientists suggests that the image was generated by some form of intense radiation emanating from the body. Such a cause would be scientifically consistent with the formation of the information for a 3D image. However, there is no physical proof to support this speculation, so the origin of the image on the Shroud remains a mystery to this day.

For centuries, the faithful have venerated the cloth as the actual burial shroud of Jesus. While significant scientific evidence now favors that idea, the Vatican still does not recognize the Shroud as a genuine relic. There are two

---

[29] Globally, individuals with AB blood type (both AB positive and AB negative) make up a small percentage of the population, typically less than 5%. AB negative is the rarest of the eight main blood types, found in roughly 1% or less of the population in most regions.

principal reasons. First, a carbon dating test of part of the shroud was conducted in 1988[30]. The result placed the origin of the cloth in the Middle Ages, between 1260 and 1390, not the time of Christ. Second, there is a lack of written evidence tracing the garment's history back to Jerusalem during the time of Christ. There is no mention of the Shroud in early Christian writings. The first official citation of it goes back to the 14th century, about the time indicated by the carbon dating test.

Yet another factor in the Vatican's decision is the anecdotal, miraculous cures attributed to the Shroud. Thus far, none of these cures have been proven to the satisfaction of the Church. The Vatican has a rigorous process for evaluating and approving reports of miraculous cures. First, the cure must be complete, instantaneous, and lasting. Second, it must be medically inexplicable. And third, it must be attributed to the involvement of the relic. None of the cures associated with the Shroud have passed this process.

The Vatican, therefore, has taken a guarded approach to the Shroud. While it has not declared the Shroud to be authentic, neither has it branded it a forgery. Instead, they refer to it as a "venerable icon." Catholics, they say, should view the Shroud of Turin with "respect and reverence", but they are not obliged to believe that it is the authentic burial cloth of Jesus Christ.

Most people, including many outside the Catholic Church, find the Shroud of Turin to be a mysterious and intriguing object, and scientific interest in its properties remains high. Some of these observers wish to continue investigating the relic. For example, scientists believe that the carbon dating test was flawed. The section of cloth chosen for the test, they say, was from a patch added to the Shroud during the Middle Ages. They want to take a piece from the central part of the Shroud and redo the test.

---

[30] **Raymond Rogers**, a chemist, found by scientific testing, that the sample used in carbon dating was likely part of the medieval repair to the Shroud. (See the detail in the update to this article)

However, even if the test showed the cloth originated at the time of Christ, the Vatican may not change its position due to other factors it considers. So, how Catholics ultimately view the Shroud of Turin will remain a matter of personal faith and devotion.

Update of the article

New Shroud dating test data - Critics of the 1988 radiocarbon testing have long maintained that the sampled section used in the examination may have been taken from a medieval patch or a contaminated area, leading to misleading results.

In 2005, Raymond N. Rogers, a chemist and Fellow of the Los Alamos National Laboratory, published a peer-reviewed paper titled *"Studies on the Radiocarbon Sample from the Shroud of Turin"* in the journal *Thermochimica Acta* (Vol. 425, Issues 1–2, pp. 189–194). His analysis focused on the *chemical and spectroscopic properties* of the area of the Shroud of Turin that was used for the 1988 radiocarbon dating, which had dated the cloth to the 13th–14th century. Rogers argued that this sample came from a *medieval repair* rather than the original linen of the shroud.

Another scientific method of measuring the age of the Shroud was implemented - A team led by Italian scientist Liberato De Caro used *the wide-angle X-ray scattering* (WAXS)[31] technique in a study of the Shroud conducted between 2019 and 2022, comparing the results to a database of linen textiles with known ages. The research culminated in a peer-reviewed study published in the journal *Heritage* on April 11, 2022.

---

[31] **Wide-Angle X-ray Scattering** (WAXS) is a powerful analytical technique used to investigate the structure of materials at the atomic or molecular scale, particularly in crystalline or semi-crystalline substances.

Unlike the more commonly known radiocarbon dating, WAXS *measures the degree of natural degradation in cellulose fibers over time by analyzing the way X-rays scatter when they pass through linen.* This method *evaluates the crystallinity of the fibers*, which decreases predictably over centuries due to environmental exposure, *providing a "clock" for dating ancient fabrics.*

The researchers' findings, published in 2022, concluded that the Shroud could *plausibly* ***date to a period between 300 BC and 400 AD,*** *distinctly in the time of Jesus, and far earlier than the* ***1988*** *carbon-**14** tests, which famously dated the cloth to between* ***1260*** *and* ***1390*** *AD.*

The team argued *that the WAXS technique is less susceptible to contamination from later repairs, handling, or biological residues, all of which could skew carbon dating results.*

These findings have reignited debates about the authenticity of the Shroud. While the WAXS study does not prove the Shroud's authenticity as the burial cloth of Jesus, it *supports the possibility that the fabric is far older than previously thought and aligns with the historical timeframe of the 1st century, the time of Jesus.*

Additional evidence suggests that the Shroud dates back to the first century in Jerusalem - The Shroud of Turin has been the subject of intense scientific and historical scrutiny for decades, with researchers seeking to determine whether it could plausibly originate from 1st-century Jerusalem. *Several lines of evidence support this claim.*

One of the most compelling comes from palynology, the study of pollen. Swiss criminologist Max Frei identified fifty-eight types of pollen on the shroud, many of which are native to the Middle East, including plants that *grow in and around Jerusalem.* This suggests that the shroud was once present in the region.

Further support comes from geological analysis. Researchers detected particles of aragonite limestone on the cloth, particularly near the foot region. This type of limestone is chemically similar to samples taken from ancient tombs in Jerusalem. Such a match strengthens the argument that *the shroud may have come into contact with a burial site in that area, consistent with the biblical account of Jesus' entombment.*

Textile experts have also *examined the linen's weave pattern, which is a 3:1 herringbone twill.* While this pattern was not typical in medieval Europe, *it is consistent with ancient weaving techniques used in the Middle East*, particularly in the regions around Syria and Palestine *during the first century*. Though this does not definitively prove the cloth's origin, it does not contradict the hypothesis of a first-century provenance and lends circumstantial support to it.

Bloodstain and image analysis add further intrigue. The blood on the cloth has been identified as type AB, which is still rare but *relatively more common among populations in the Middle East.* Forensic studies on the Shroud indicate that the blood shows signs of clotting and trauma consistent with crucifixion wounds.

In summary, the Wide-Angle X-ray Scattering experiment conducted by Liberato De Caro in 2022, along with the chemical analyses, including spectroscopic tests, performed by Raymond Rogers in 2005, suggest an older date for the linen fibers, consistent with the 1st century. ***This evidence supports the idea that the Shroud of Turin may have originated in the Jerusalem area around the time of Jesus.***

How was the image formed on the Shroud of Turin? - One of the more scientifically intriguing theories about the formation of the image on the Shroud of Turin suggests that *it was caused by a burst of radiation emanating from the body wrapped within the cloth.* This hypothesis aims to explain several of the Shroud's unique characteristics, particularly the superficiality of the image (affecting only the top 1–2 fibrils of the linen threads), its photonegative quality, and the presence of three-dimensional, spatial data embedded in the image intensity.

Proponents of this theory argue that *no known artistic method or natural decay process can account for these features, and therefore, an unusual burst of energy, possibly at the moment of resurrection, may have imprinted the image.*

Now, there is no known scientific mechanism for a human body to emit high-energy radiation spontaneously. To speculate on this as the mechanism that formed the image on the Shroud is to assume a miracle took place.

Among the different forms of radiation proposed, ultraviolet (UV) radiation has been the subject of experimental testing. In 2010–2011, a team led by Dr. Paolo Di Lazzaro at Italy's ENEA research agency demonstrated that a short, intense burst of UV light from an *excimer laser* could *create surface discoloration on linen that mimics the Shroud's image.*

Their results showed that such radiation could alter only the top 200 nanometers of linen fibrils, aligning with the superficial nature of the Shroud image. However, *the amount of energy required to produce this effect would have to be extraordinarily high, several billion watts in an extremely short burst, well beyond any known natural source, and certainly beyond what a human body could produce.*

Other versions of the radiation hypothesis involve subatomic particles. Some researchers, such as Dr. John Jackson of the Shroud of Turin Research Project (STURP), have proposed that *a burst of proton radiation from the body* could have interacted with the linen, causing oxidation and dehydration of the cellulose that led to the formation of the image.

Similarly, neutron radiation has been proposed by others, including physicist Giulio Fanti. Neutrons could cause nuclear reactions within the elements of the linen fibers, producing the image chemically.

Both particle radiation theories, however, suffer from the same core problem as the UV theory. There *is no known biological mechanism that can account for such emissions, especially in a burst strong enough to cause the observed effects.*

Some more speculative theories even *propose photon emissions, perhaps soft X-rays or visible light,* radiating from the body in a controlled or coherent fashion. *These theories verge into the metaphysical by implying a resurrection event that triggered an unknown release of energy.* While they attempt to account for the image's properties without requiring contact or pigments, they are still untestable and outside the scope of current physical science.

In summary, the radiation theory of image formation offers an explanation that aligns closely with the Shroud's physical and chemical properties, including the extraordinary superficiality, lack of pigments, and the realism of the image. It remains one of the few scientific hypotheses that attempts to reconcile the cloth's mysteries without resorting to artistic forgery. However, it also raises profound questions about the nature of the energy involved, its source, and *whether the event that created the image lies within or beyond the boundaries of known physics.*

The conclusions of some scientists about the authenticity of the Shroud of Turin – The Shroud of Turin, a linen cloth bearing the faint image of a man, is believed by many to be the burial shroud of Jesus of Nazareth. While mainstream science has generally remained skeptical, *several scientists over the decades have argued in favor of the Shroud's authenticity* based on what they see as compelling scientific evidence. These researchers often come from diverse disciplines, including physics, chemistry, forensic pathology, and image analysis. Below are *several key scientists who believe in the authenticity of the Shroud, along with their reasons based on the evidence.*

Dr. John Jackson – Physicist and STURP Leader

Dr. John Jackson, a physicist from the U.S. Air Force Academy, was one of the leaders of the Shroud of Turin Research Project (STURP), which conducted the most comprehensive scientific analysis of the Shroud in 1978. *Jackson's work involved studying the image's unique three-dimensional qualities using image-processing technology. He found that the image was* ***not*** *made by paint, dye, or scorching,* and argued that *it contains distance-coded information, meaning the image intensity varies with the distance between the cloth and the body.* **This phenomenon is not reproducible by any known artistic method**. Jackson and his colleagues proposed that *the image might have been formed by a burst of radiant energy emanating from the body, consistent with a supernatural event such as the Resurrection.*

Dr. Alan Whanger – Radiologist

Dr. Alan Whanger, a radiologist at Duke University, studied the Shroud using polarized image overlay techniques.[32] He claimed to have *found congruencies between the Shroud and ancient coin images from the first century, particularly coins known to have been minted during the reign of Pontius Pilate.* ***The presence of these coins adds to the evidence for the Shroud originating in first-century Jerusalem.***

Whanger also noted anatomical precision and wound patterns on the Shroud that he argued matched the Gospel descriptions of Jesus' crucifixion. He suggested that such details would have been unlikely to be known or reproduced by a medieval forger. Although some have criticized Whanger's methods as subjective, he *maintained that the cumulative evidence pointed strongly to authenticity.*

Dr. Giulio Fanti – Mechanical Engineer

Professor Giulio Fanti of the University of Padua in Italy has conducted extensive research on the Shroud's physical and chemical properties. In particular, Fanti *used infrared spectroscopy, Raman spectroscopy, and mechanical strength tests on linen fibers.* ***His tests indicated that the Shroud linen dates to between 300 BC and 400 AD***, contradicting the famous 1988 radiocarbon dating that dated it to the Middle Ages (1260–1390 AD). Fanti has argued that the 1988

[32] **Polarized Image Overlay Techniques** involve the use of polarized light to combine two or more images in a way that enhances or isolates certain visual information. This technique takes advantage of light polarization — the orientation of light waves — to control how images are captured, processed, and displayed.

dating was skewed due to contamination or the tested sample coming from a repaired area of the cloth. He has also proposed that *the image was created by an intense burst of energy, possibly ultraviolet radiation, which again aligns with theories of a supernatural event.*

Dr. Pierre Barbet – Forensic Pathologist

Dr. Pierre Barbet, a French surgeon and forensic expert, conducted a series *of medical studies* on the Shroud beginning in the 1930s. *He examined the bloodstains and wound patterns and concluded they were anatomically accurate and consistent with Roman crucifixion practices.* Barbet's most famous work, *A Doctor at Calvary*, argued that an artist could not have produced the image and bloodstains and that the man of the Shroud had endured injuries that matched the Gospel descriptions of Jesus' Passion, including scourging, crowning with thorns, and a side wound. *His medical expertise led him to believe that only a real crucifixion victim could have produced such forensic details.*

Dr. Raymond N. Rogers – Chemist (Cautious Support)

Raymond N. Rogers, a chemist from Los Alamos National Laboratory and part of STURP, *initially accepted the 1988 radiocarbon dating results.* However, *later in his career, he revisited the issue and concluded that the carbon-dated sample was from a re-woven area, not original to the cloth.* Using chemical tests, Rogers found differences in the vanillin content of the tested sample compared to the rest of the cloth. Vanillin, a degradation product of lignin in flax fibers, disappears over time. Its presence in the tested sample but absence elsewhere led him to propose that the tested fibers were much younger than the original Shroud linen. *Though Rogers stopped short of affirming miraculous origins, he became convinced that the cloth itself is older than the radiocarbon tests suggested.*

Summary

While none of these scientists claimed absolute proof, they all believed the physical, chemical, anatomical, and image characteristics of the Shroud pointed strongly toward its authenticity. Their arguments are often centered around the

Shroud's inexplicable image formation, the forensic consistency with crucifixion, and challenges to the 1988 carbon dating results. Nevertheless, the debate continues, as the Shroud straddles the complex boundary between science, history, and faith.

Intelligent Design: Can scientists find God in nature?

First published in the Gloucester Daily Times on June 14, 2023

Most scientists who were active in the nineteenth century considered themselves religious and believed in God. That all changed with Darwin's publication of **The Origin of Species by Means of Natural Selection** in 1859. Suddenly, scientists had an explanation for creation that didn't depend on the existence of God. *A 2014 study by the Pew Research Center found that about 80% of the scientists in the U.S. claimed to be either atheists or agnostics.*

However, some scientists still believe that there is an intelligent cause in the formation of the universe. A primary reason for their belief is that Darwin's theory falls short in explaining some important aspects of the history of life on Earth. This is especially true for the creation of the living cell.

Dr. Garry Nolan is a highly regarded professor at Stanford University's School of Medicine. During a recent interview, he was asked, "What is the most beautiful or fascinating aspect of human biology at the level of the cell?"

Without hesitation, he replied: "The micro-machines…that proteins make and become (under the direction of the digital information found in the DNA)….If you ever wanted to believe in God, just look inside the cell (and see how it functions)."

The cell Dr. Nolan refers to is the smallest part of living matter. Cells are the specialized building blocks of multicellular organisms, including plants and animals. Not until the 1930s, long after Darwin's death, did scientists begin using the newly invented electron microscope to examine the interior of a cell. They discovered a vast array of protein structures, or, as Dr. Nolan referred to them, *molecular machines*. These machines perform all the functions necessary to keep the cell alive and functioning properly.

Dr. Bruce Alberts, a former President of the National Academy of Sciences, expanded on the awesome aspects of the cell's inner workings. He wrote a 1998 article for "Cell", one of the world's foremost biology journals. In the piece, he said: "The entire cell can be viewed as a factory that contains an elaborate network of interlocking assembly lines, each of which is composed of a set of large protein machines. Like machines invented by humans to deal efficiently with the macroscopic world, these protein assemblies contain highly coordinated moving parts." Alberts went on to praise the "speed," "elegance," "sophistication," and "highly organized activity" of the "remarkable" and "marvelous" structures inside the cell.

But how did the intricate, massively complex molecular machines that make up the cell come into being? Today, Evolutionary Biologists believe that the existence of cellular machines is the result of the processes described by Darwin in his theory of evolution by natural selection. *These machines, they say, represent the end products of a long series of random modifications to successive generations over billions of years.* The scientists state that it looks like "design," but it's just Evolution in action.

Darwin's Theory of Evolution is well-accepted by scientists because of the large amount of evidence that proves it is a good representation of reality. *However, it does have some serious, well-recognized weaknesses.* One of the most profound is that the design of the cell, including its very intricate DNA code, is the result of Evolution. Believers in Intelligent Design argue that it would be impossible for such a large number of sophisticated, complex machines to have evolved through the random, blind, undirected processes described by Darwin. They think that there had to be an "intelligent designer" involved, *particularly because these molecular machines are "irreducibly complex".*

Dr. Michael Behe, a Professor of Biological Sciences at Lehigh University, was one of the first scientists to talk about a cell's "irreducibly complex" components. He said that a molecular machine is a "single system" comprised of multiple, interacting mechanisms that are all necessary for it to function. To visualize his point, consider that a mousetrap is a macro example of such a "single system". If any one of the components of the mousetrap (the base, hammer, spring, catch, or holding bar) is removed, then the trap doesn't function. And the same holds for the cell.

Behe argues that *many custom-designed pieces are all required to fit together before a molecular machine's function appears.* A predecessor machine would have had some non-functioning parts and could not have evolved gradually into a complete operating structure. *Behe and other scientists conclude, therefore, that it had to have been designed by an intelligent agent.*

When Michael Behe presented his "irreducibly complex" argument, many other scientists, who were strong proponents of Darwinian evolution, blasted his statements. However, the evidence supporting their positions appears to be weak.

Microbiologist James Shapiro of the University of Chicago, who is a member of the Darwinist community, declared in a National Review article that *"There are no detailed, Darwinian accounts for the evolution of any fundamental biochemical or cellular system, only a variety of wishful speculations."*

Mainstream scientists still do not accept Intelligent Design as a valid area of scientific investigation. However, those who support the possibility, such as The Discovery Institute based in Seattle, Washington, present their case in numerous peer-reviewed articles, books, presentations, podcasts, and professional meetings. They look at the universe, the world, and its creatures, and, like Dr. Garry Nolan of Stanford University, see the hand of God in all of it.

Update of the article

Recent thinking on Intelligent Design - In recent years, proponents of intelligent design (ID) have continued to develop arguments and present new research aimed at demonstrating that the structure and complexity of the universe, life, and biological systems point to the activity of an intelligent cause rather than undirected natural processes.

One of the most active areas of development has been in the field of molecular biology, where researchers highlight the intricate machinery within living cells, *particularly the information-rich structures of DNA and the highly regulated processes of gene expression,* as evidence of design.

Advocates argue that the digital coding properties of DNA resemble human-designed information systems, which, they claim, cannot be explained by random mutation and natural selection alone.

Dr. Stephen C. Meyer, a leading figure in the ID movement, elaborates on this in his 2021 book *The Return of the God Hypothesis*, where he contends that *the origin of biological information, the fine-tuning of the physical constants of the universe, and the intelligibility of the natural world* are best explained by a transcendent, designing intelligence.

Another significant development in the intelligent design field is the expanding study of ***fine-tuning*** *in cosmology.* Many physicists and philosophers, including those outside the ID movement, acknowledge that *the fundamental constants of nature (such as the gravitational constant, the cosmological constant, and the ratio of the strengths of fundamental forces) appear to be finely tuned to an extraordinary degree.* ***If any of these values were even slightly different, life as we know it would not be possible.*** ID theorists argue that this fine-tuning strongly suggests a purposeful calibration by an intelligent agent, since the odds of such life-permitting values arising by chance are astronomically low.

Some scientists invoke the multiverse[33] as an alternative explanation. In this theory, there are an infinite number of universes, and we happen to live in one that has precisely the correct set of constants to allow the world we know.

ID proponents counter that the multiverse hypothesis is speculative and lacks empirical support, making design a more parsimonious inference based on the available evidence.

ID researchers have also leveraged advances in information theory and systems engineering to challenge naturalistic explanations for complex biological systems. The concept of *irreducible complexity*—first popularized by biochemist Michael Behe—continues to be applied to cellular mechanisms like the bacterial flagellum, the blood-clotting cascade, and protein assembly lines. *Behe and others argue that such systems require all their parts to function and could not have evolved gradually through step-by-step Darwinian processes.*

Critics of ID dispute this, claiming that evolutionary pathways are plausible or already known for many such systems. However, ID advocates maintain that no known unguided mechanism has adequately explained how the necessary information and integration arise spontaneously.

While intelligent design remains controversial in mainstream scientific circles, mainly due to its perceived theological implications and lack of testable predictions, it continues to gain traction in interdisciplinary discussions among scientists, philosophers, and theologians. ID thinkers are increasingly framing their arguments, not as gaps in scientific understanding (a "God of the gaps" argument), but rather as favorable inferences based on empirical evidence for design. The debate is no longer confined to biology but has expanded to include physics, cosmology, and even

---

33 The **multiverse** is a theoretical concept in physics and cosmology suggesting that our universe is just one of many — perhaps infinitely many — universes that exist in a larger "multiverse." Each of these universes may have different physical laws, constants, dimensions, and histories.

consciousness studies, as researchers explore whether the apparent purpose, order, and intelligibility of the universe suggest an underlying intelligent source.

Regenerative medicine is coming of age

First published in the Gloucester Daily Times on July 6, 2023

The ability to regenerate body parts is a common phenomenon in the animal kingdom. We see some examples in our Gloucester waters. Starfish can regenerate lost limbs, as can horseshoe crabs, jellyfish, sea urchins, and sea cucumbers. In our Dog Town forest, white-tail bucks regenerate their antlers every year. The antlers are made of bone, nerves, and skin and are the fastest-growing tissues in the animal kingdom, up to one inch per day!

Humans are also capable of regenerating specific tissues. The liver can regenerate lost or damaged parts as long as a minimum amount remains intact. Many people don't realize that young children can regenerate fingertips that may have been lost in an accident. This ability disappears, however, at about the age of eight or nine.

For generations, the dream of scientists has been to understand how this regenerative ability works in animals and to find a way to harness it for humans. In fact, interest in Regenerative Medicine dates back to the nineteenth century. That's when the idea was conceived for grafting skin from one part of the body to another.

Then, in the 1970s, stem cell research began. *Stem cells are undifferentiated cells that can self-renew and differentiate into any cell in the body.* So, it appeared that a door had opened. However, progress has been slow due to several safety concerns that have emerged. It was discovered that stem cells could potentially lead to the development of uncontrolled tumors. An additional problem was that the body could reject them if they originated in another person.

Recently, however, significant progress has been made in regenerative medicine. In fact, it is the innovative work of a North Shore scientist that is bringing us closer to the dream of complete body part regeneration in humans. Professor Michael Levin of Tufts University in Medford is the head of the Center for Regenerative and Developmental Biology. He and his colleagues have made some extraordinary progress with groundbreaking ideas and approaches.

Michael was born in Moscow, Russia, but grew up in the Lynn area and now resides in Beverly. *Many of Levin's breakthroughs came from his study of Planaria, an immortal (yes, really!) half-inch-long flatworm that lives in freshwater habitats around the world.*

The Planaria are immortal because they can regenerate their entire bodies from any fragment. They are believed to be approximately 400 million years old and do not appear to age in the same manner as humans. Their cells don't accumulate damage over time, and they don't show signs of decline in their ability to regenerate. Cut them in half, and the part with the head will grow a new tail, and the part with the tail will grow a new head, including the brain and eyes. Scientists have cut a single specimen into a record 279 segments, and each one regenerated into another complete copy of the same individual. *And it has been proven through Levin's experiments that the creatures even retain the memories of their precursors.*

Professor Levin has spent many years studying how the Planarians accomplish their "magic". *He found that the ability of Planarians to regenerate is due to the presence throughout their bodies of a type of stem cell, called a neoblast.* Like all stem cells, they are undifferentiated and can self-renew, as well as differentiate into any cell in the body.

In a breakthrough, Levin has discovered that planarians possess a complex system of bioelectric signaling pathways that coordinate the activity of their neoblasts. These signaling pathways go beyond the information provided in the creature's DNA. *Instead, these bioelectric signals are responsible for directing the neoblasts to the site of injury, instructing them to differentiate into the correct cell type, and then stopping when regeneration is complete.*

Levin calls this the "bioelectric code", believing it is key to the dream of regenerating human body parts. Essentially, bioelectricity acts as a master switch to turn on the chemical and molecular signals that create the new structure. Scientists don't have to get involved with the actual regeneration since the body takes care of that.

Using this relatively simple approach, the group led by Levin at Tufts has had success in regenerating the limbs of African clawed frogs. These species typically do not regenerate limbs. A drug cocktail to activate the regeneration master switch was applied for 24 hours with a wearable bioreactor attached to the stump of the amputated limb. With this treatment, and after eighteen months, the frog's body restored a complete leg with all its sensitivity and functionality. This was the first time Regenerative Medicine researchers have ever accomplished such a feat.

Along with his colleague, David Kaplan, also a Tufts professor, Levin has formed a start-up company to commercialize this technology. The company, Morphoceuticals, Inc., is in Marblehead. The company's goal is to harness its bioelectric code technology and become the first in the world to regenerate fully functional limbs in mammals. They are now working with mice to make it happen. And when it does, it will be a breakthrough in the field of Regenerative Medicine.

The dream of regenerating human limbs and vital organs is now in sight. Many people who are hurting can't wait for the reality to be achieved. Levin is 54 and has stated that he firmly believes it will happen in his lifetime. Hopefully, most of us will also be around when he succeeds.

Update of the article

Additional areas of advanced biological research - Xenobots, also known as biobots, are an innovative creation from Professor Michael Levin's lab at Tufts University, developed in collaboration with computer scientist Josh Bongard. These *tiny living robots* are constructed from stem cells derived from the African clawed frog (*Xenopus laevis*), hence the name.

Remarkably, the cells are reprogrammed and reshaped, not genetically altered, into *entirely new configurations that can move, work collectively, self-heal, and adapt to their environment.* They represent an entirely *novel class of life form*: neither traditional robots nor naturally occurring organisms, *but programmable biological machines.*

The scientific significance of Xenobots is multifold. First, they provide a unique window into understanding how individual cells collaborate to form larger, functional structures, a crucial question in developmental biology and regenerative medicine.

Second, they challenge our understanding of what it means to be "alive" and *opens the possibility of using engineered living systems for practical purposes.* Already, Xenobots have been shown to move autonomously, push microscopic objects, and even exhibit basic forms of memory.

Looking ahead, researchers envision using them for a *range of applications,* including *targeted drug delivery, environmental cleanup (such as microplastic collection), and tissue repair.* The long-term outlook is auspicious but cautious. Xenobots are still in early experimental stages, and many scientific, ethical, and safety questions remain. However, their potential to revolutionize both medicine and synthetic biology is profound.

We are close to reaching life expectancy "escape velocity"

First published in the Gloucester Daily Times on February 8, 2023

When I was born in 1939, the statistical life expectancy for people was 62 years. This would remain my fate if nothing changed. But over the ensuing decades, science made great strides in many fields, all of which contributed to increasing humans' prospects for a longer life. A baby born today can expect to live to be 79. That's seventeen years longer than when I was born.

Due to the same advances in science and technology, my life expectancy increased every year I lived. I have received sophisticated medical treatments and take a range of advanced medications that keep my diseases at bay. Most of these innovative therapies were not even available thirty or forty years ago. According to an online life-expectancy calculator that takes into account my situation and lifestyle, my life expectancy is now 91.

Until recently, the increase in human longevity has been changing at a linear rate, approximately 2 years per decade. However, due to numerous recent breakthroughs in various fields, the increase in longevity is poised to shift into an exponential rate of change.

When it reaches the point where, for each year that you live, your life expectancy increases by more than a year, humans approach the longevity "escape velocity". Life expectancy will then have increased more than ten years per decade. And, at that point, 100 years old will become the new 60.

Ray Kurzweil, a world-renowned inventor and futurist based in Newton, forecasts that we will reach longevity "escape velocity" by the end of this decade. He may be optimistic, but he believes that a convergence of significant developments in many fields will contribute to this extension of the human lifespan. The technologies he is considering include artificial intelligence, cloud computing, networks, sensors, robotics, massive datasets, biotechnology, and nanotechnology.

An example of a *biotechnology development that could be very significant in extending life will be the availability of 3-D printed organ parts.* It sounds like science fiction, but we may see some examples hitting the market this year. Organovo, a California-based company, was the first to successfully engineer commercially available, 3D-bioprinted, human livers and kidneys. In fact, some scientists are saying that, by 2028, death by organ failure may become a problem of the past.

While Dr. David Sinclair may not share Ray Kurzweil's level of optimism about reaching longevity escape velocity, he does believe that the first person to reach the age of 150 is alive today. Sinclair is the Director of Harvard Medical School's Paul F. Glenn Center for Biology of Aging Research. He and his associates are working on a particular biotechnology development that will contribute to achieving that goal.

Up until recently, normal aging has been assumed to be associated with the accumulated mutations, or unexpected changes, to a human's DNA. That's the molecule that carries all our genetic information. These mutations gradually disrupt the body's normal functioning. And organs and other body parts deteriorate to the point where their regular roles cease.

Even with today's advanced technologies, it is challenging to alter DNA to the extent that all DNA mutations can be halted or reversed. You would need to repair trillions of altered genes to reverse the effects of aging. Fortunately, Dr. Sinclair and his associates from Harvard Medical School found a way around that problem.

They discovered that they could reverse the aging clock in laboratory mice by altering the animal's "epigenome". This is a suite of molecules that turn the mice's DNA on and off in different body tissues. Altering the epigenome is much easier than altering DNA itself.

Sinclair already has an epigenome-altering drug in development that resets the age of a monkey's eye to cure its blindness. He believes that this same technology can be used to reset the age of any part of the body. That includes the brain. In tests on mice, after treatment with this drug, very old individual rodents were able to learn again.

Sinclair hopes that his team will ultimately be able to reset the human body's clock, allowing it to fight diseases whose incidence increases with age. These include cancer, heart disease, and Alzheimer's. It would represent an entirely new approach to treating the most serious diseases affecting humankind. Individuals would also look and feel much younger.

The Harvard team is optimistic that they will succeed in developing this technology over the next several years. Their primate studies look promising, and human studies will follow soon after the tests on the monkeys are completed.

Due to Sinclair's publications in medical journals, many other researchers are pursuing similar research tracks. Sinclair estimates that approximately $5 billion has been invested recently in the development of anti-aging drugs. Therefore, he is confident that someone will soon succeed in proving the efficacy of the Harvard approach in humans.

One of Sinclair's close associates at the Medical School, Jae-Hyun Yang, said, "These technologies are developing now, and the speed of development is getting faster and faster. So, I don't think it's far away that people will live to 150."

Update of the article

Recent developments in the 3D printing of human organs -

Over the past two years, the field of 3D-printed organ replacements has made significant strides, bringing the prospect of lab-grown organs closer to clinical reality. While fully functional, transplantable organs are not yet available to the public, recent advancements in bioprinting technology have led to promising developments in tissue engineering and organ scaffolding.[34]

In 2024, researchers achieved a notable breakthrough by successfully bioprinting vascularized tissues, including components of the heart, liver, and kidney. These advancements address the critical challenge of integrating blood vessels into bioprinted tissues, which is crucial for sustaining organ function after transplantation. Additionally, a team at Northeastern University developed a novel elastic, hydrogel material designed for 3D printing soft living tissues, paving the way for more durable and functional organ constructs.

Government initiatives have also propelled the field forward. The Advanced Research Projects Agency for Health (ARPA-H) launched the PRINT program in 2024, aiming to develop on-demand bioprinting processes for vital organs such as the kidney, heart, and liver. This program focuses on leveraging patient-derived cells to create immune-matched organ replacements, thereby potentially reducing the risk of transplant rejection.

Private companies are making significant contributions as well. United Therapeutics has been at the forefront, developing complex 3D-printed lung scaffolds containing

---

[34] **Organ scaffolding** is a technique used in tissue engineering and regenerative medicine to create a **3D structural framework** that mimics the architecture of a human organ. This scaffold acts as a **support system** for cells to grow, organize, and eventually form functional tissue or entire organs.

trillions of voxels[35] and extensive networks of capillaries and alveoli[36]. These scaffolds are designed to mimic the intricate structure of human lungs, representing a substantial step toward functional organ replacements.

While these advancements are promising, *experts caution that widespread clinical application of fully functional 3D-printed organs remains years away.* Ongoing research and clinical trials are crucial for addressing challenges related to organ functionality, integration, and long-term viability. Nonetheless, the rapid progress in bioprinting technology offers hope for addressing the global organ shortage crisis in the foreseeable future.

David Sinclair's continuing research on anti-aging medications -

Harvard Medical School geneticist Dr. David Sinclair and his team continue to push the boundaries of anti-aging research, unveiling several promising approaches that could fundamentally reshape our understanding and treatment of aging.

Another breakthrough, after his epigenome studies, involves the identification of six small-molecule cocktails that can reverse cellular aging in human skin and muscle cells within days. These molecules, unlike gene therapies, could potentially be delivered as oral medications, opening the door to a future where age-related decline is treated like a chronic condition. While these results are still preliminary, they offer strong support for Sinclair's assertion that biological aging can be targeted and even reversed with the right interventions.

---

[35] **Voxels** define the **precise placement of cells or materials** in a 3D-printed tissue or organ. Millions or even trillions of voxels may be used to design complex biological structures like lungs or kidneys.

[36] **Alveoli** are tiny, air-filled sacs in the lungs where **gas exchange** occurs between the air we breathe and the blood circulating in the lungs. They are the final, smallest branches of the respiratory system and are crucial for oxygenating the blood and removing carbon dioxide.

Sinclair is also known for his advocacy of NAD+ precursors[37] such as NMN (nicotinamide mononucleotide), compounds that fuel cellular repair mechanisms and have shown lifespan-extending effects in mice. *Recent, unpublished studies from his lab suggest that NMN may also reduce frailty* and provide protection against radiation. Although such claims remain under scientific review, Sinclair continues to incorporate these compounds into his own health regimen, alongside resveratrol, metformin, spermidine, and other supplements intended to slow the aging process.

Looking ahead, *clinical trials led by Sinclair-affiliated companies are expected to begin as early as 2025.* One such effort, through *Life Biosciences,* involves evaluating a gene therapy called ER-100 for optic neuropathies. If successful, this could represent the first practical step toward reversing specific aspects of aging in humans. Sinclair believes that, within five years, small-molecule therapies could be on the market—*pills that activate the body's rejuvenation pathways affordably and safely.*

Despite growing excitement, Sinclair's work has not been without controversy. Critics have raised concerns over the commercial push behind some of his ventures and the *need for more rigorous peer-reviewed validation of his lab's findings.* Still, his vision for the future of medicine is clear: aging, once thought to be inevitable, may soon be optional. Whether these breakthroughs live up to their promise remains to be seen, but Sinclair's research is undeniably shifting the scientific conversation from slowing aging to reversing it.

---

[37] **NAD+ precursors** are molecules that the body uses to produce **nicotinamide adenine dinucleotide (NAD^+)**, a vital coenzyme found in all living cells. NAD^+ plays a key role in cellular energy production, DNA repair, and regulating many biological processes tied to aging and health.

The day I flatlined

First published in the Gloucester Daily Times on September 2, 2019

I was in a recovery room at Lahey Hospital after having had a routine heart diagnostic procedure. The last thing I remember was sitting up in bed, checking my cellphone. The nurse was nearby, at her desk, completing some paperwork. Then everything went black. I had a sudden cardiac arrest. This means that my heart had abruptly and unexpectedly stopped beating. As a result, my blood stopped flowing, I stopped breathing, my pupils dilated, and I lost consciousness within seconds. I had flatlined and, for all intents and purposes, I was dead.

Currently, a patient who suffers a sudden cardiac arrest in a hospital has about a twenty-two percent chance of being resuscitated and surviving to discharge. That means that four out of five patients who experience such an event don't make it. Pretty bad odds! And it is much worse outside a hospital. With no expert medical personnel or an Automatic External Defibrillator close at hand, less than 10% of sudden cardiac arrest patients survive.

You may remember a movie made in 1990, remade in 2017, called "Flatliners". In that film, five medical students experiment using drugs to induce a sudden cardiac arrest purposely. Their idea was to have a "near-death" experience and get a glimpse of the afterlife before their friends revived them.

But, in real-life, not all patients who survive their cardiac arrest have a near-death experience. It's only about one in five of them that do.

Over the years, however, there have been hundreds, if not thousands, of anecdotal reports of near-death experiences from many countries and cultures. *The similarities in these stories sparked the interest of the scientific community, and as a result, several rigorous scientific studies of the phenomenon have been conducted.*

For example, the renowned, international medical journal, *Lancet*, published a study of near-death experiences by patients with sudden cardiac arrest. It was conducted by a Dutch cardiologist, Pim van Lommel.

The experiences of his patients tended to be like many of the other recorded near-death experiences found in the literature. They often entail being conscious, but knowing you are dead, and finding yourself in a dark tunnel with a bright light at the other end. When the subject emerges into the light, they find themselves in a most beautiful place and have feelings of total serenity, security, warmth, and unconditional love.

Many experiencers also see long-dead relatives, friends, and even pets. The patient can interact with them, and it all seems very real and comforting. Many of these patients claim that their experience changed their lives tangibly and positively afterwards.

Other sudden cardiac arrest patients claim to have a different kind of incident. It is an out-of-body experience that also feels very real to them. In these cases, the patient believes that they are floating above their hospital bed, looking down on the scene where the medical staff is trying to resuscitate them. Many patients can recount the actual actions and conversations that took place, confirmed by the participants. Yet, they had flatlined, their brains were completely inactive, and they had no senses.

Many doctors and scientists have tried to develop a reasonable medical explanation for these experiences. Various theories have been proposed, but there is no scientific consensus on how these events can occur.

One of the most interesting aspects of near-death experiences is the idea of "consciousness" and how it relates to this mysterious phenomenon. Most dictionaries define consciousness as "the fact of awareness by the mind, of itself and the world." But, when blood stops flowing to the brain because of cardiac arrest, the victim's brain activity ceases, and they quickly lose consciousness. Yet, these victims still claim to have such conscious experiences.

Van Lommel concluded in his *Lancet* article that the results of his study demonstrate that the patient's awareness indeed continued after all brain activity ceased. In his summary, he raised the question, *"If consciousness is a function of brain activity, as most scientists believe, how could consciousness be experienced after the brain no longer functions?"*

Van Lommel answered his question in a later book that he wrote. It was titled **Eternal Consciousness**. He claimed in the book, like some philosophers have in the past, that human consciousness functions independently of the brain. The brain, he says, "is just the receiver of the signals generated by this (higher order) consciousness, much as a television is the receiver of signals from a television station." This concept, some philosophers say, is evidence of the existence of the soul, and the near-death experience provides a glimpse of the afterlife. In the *Flatliners* movie, it is this glimpse that the student doctors were seeking.

Other doctors, trying to explain the conscious experiences of cardiac arrest patients, attribute it to the act of heart massage in the resuscitation process. The result of the massage, they say, is that there is some blood flow to the brain. And that's what explains the consciousness required for the near-death experiences.

On the other hand, van Lommel's view is supported by many neurologists. They say that until the heart starts beating again, there can be no brain function.

There are, however, a few neurologists who support the idea that consciousness operates independently of the brain. Dr. Donald Hoffman, Professor of Cognitive Science at the University of California, Irvine, hypothesizes exactly that idea. He developed a mathematical, scientific analysis to prove it.[38]

In my case, I am obviously among the one in five patients who suffer a sudden cardiac arrest in a hospital and survive. After everything went black, the next thing I knew, I was

---

[38] See the Update of this article for a discussion of Hoffman's theories.

looking up at a bearded male nurse, whom I later learned was named Eddie.

Eddie and the other medical professionals in that recovery area, went into quick action and brought me back within a few minutes. As I returned to full consciousness, I remember the crowd around me and the frantic looks on their faces, but nothing more.

The doctor in charge of my case called my wife to inform her of what had happened. When she arrived to see me at the hospital, I told her how disappointed I was. I had a sudden cardiac arrest and did not find myself in a dark tunnel or come out of it into the bright light to find myself in a beautiful place. Or see any relatives or pets.

Now, that would have been a great story to be able to tell. Maybe next time.

Update of the article

Dr. Pim van Lommel's recent publications on his theory regarding near-death experiences - Van Lommel has continued to contribute to the fields of near-death experiences (NDEs) and consciousness studies over the past five years. His recent work reinforces his longstanding hypothesis that consciousness may not be solely a product of brain activity and could persist independently of it.

In 2021, Dr. van Lommel authored an essay titled *The Continuity of Consciousness,* which was awarded second prize in the Bigelow Institute for Consciousness Studies' international essay contest. In this work, he argues that consciousness is non-local and continuous, not confined to the brain or body. He supports this claim with empirical data from near-death experiences (NDEs), particularly those occurring during cardiac arrest when brain activity is absent. Van Lommel posits that the brain functions more as a facilitator of consciousness rather than its originator.

In 2022, he co-authored a commentary in *Frontiers in Aging Neuroscience* discussing a study on electroencephalogram

(EEG) activity in a dying human brain. Van Lommel and his colleagues questioned the interpretation that such EEG patterns indicate residual consciousness, *suggesting instead that these findings might support the notion of consciousness existing independently of brain function.*

Furthering this discourse, in 2023, van Lommel co-authored another commentary in the *Journal of Near-Death Studies*, critically evaluating reports of electrical activity in the human brain during the dying process. He emphasized the need for careful interpretation of such data, cautioning against the assumption that equates neural activity with conscious experience.

Beyond academic publications, Dr. van Lommel has actively engaged in public discussions to disseminate his findings. In May 2022, he participated in a podcast with Buddhist monk Geshe Tenzin Namdak, exploring the intersections between near-death experiences (NDEs), consciousness, and Buddhist philosophy. They discussed how experiences reported during near-death experiences (NDEs) might align with concepts of consciousness found in contemplative traditions.

In July 2023, he was featured in an interview with the *Essentia Foundation*, where he delved into the philosophical implications of his research on consciousness beyond death. He reiterated his stance that NDEs challenge the conventional materialistic view of consciousness, *suggesting instead a model where consciousness is fundamental and not merely a byproduct of brain activity.*

Dr. van Lommel's recent work continues to influence both scientific and public perspectives on consciousness and NDEs. His research supports the idea that consciousness may be a non-local phenomenon, not entirely dependent on brain function. Through his publications and public engagements, he advocates for a broader understanding of consciousness that encompasses experiences reported during near-death experiences (NDEs), challenging traditional materialistic paradigms.

Dr. Donald Hoffman's position on consciousness operating independently of the brain -

Dr. Donald Hoffman, a cognitive scientist at the University of California, Irvine, has continued to advance his theories on consciousness over the past five years, challenging conventional views in neuroscience and physics. His recent work further develops his *"conscious agent" framework*, proposing that *consciousness is fundamental and that the physical world emerges from it, rather than the other way around.*

In his 2023 paper, *Fusions of Consciousness*, Hoffman and colleagues present a mathematical model describing how "conscious agents", entities with the capacity for experience and decision-making, interact and combine. They argue that these agents exist outside spacetime and that their interactions give rise to the physical phenomena we observe. The model employs Markov chains to represent the dynamics of these agents, *suggesting that what we perceive as the physical world is a projection of these deeper, conscious processes.*

Further elaborating on this framework, the 2024 article *"Interfacing Consciousness" discusses how spacetime and physical objects are not fundamental realities, but rather emergent constructs arising from the interactions of conscious agents.* This perspective aligns with views in theoretical physics that question the fundamentality of spacetime, proposing instead that consciousness itself forms the basis of reality.

Dr. Hoffman has actively engaged in public discourse to disseminate his ideas. In a 2023 presentation titled *Consciousness and Its Physical Headset* at the **Consciousness & Reality Colloquium**, he argued that traditional theories placing consciousness as a byproduct of physical processes are inadequate. Instead, he *posits that consciousness is primary and that the physical world is a kind of "headset" or interface through which conscious agents interact.*

Additionally, in a 2023 podcast interview, Hoffman discussed his view that *our perceptions are akin to a user interface, evolved not to reveal objective reality but to guide adaptive behavior*. He suggests that what lies beneath this interface is a realm of interacting conscious agents, and that understanding this realm is key to understanding consciousness itself.

Core Concepts of Dr. Hoffman's Theory:

- *Conscious Agents*: Fundamental units of consciousness that interact outside of spacetime, forming the basis of reality.
- *Interface Theory of Perception*: The idea that our perceptions are not accurate representations of the world but evolved interfaces that guide behavior.
- *Spacetime as Emergent*: The proposition that spacetime and physical objects are not fundamental but arise from the interactions of conscious agents.

Dr. Hoffman's work continues to provoke discussion and debate, offering a radical rethinking of consciousness and its place in the fabric of reality.

How Technology reveals the brain's secrets

Submitted to the Gloucester Daily Times but not yet published.

Many of us are fascinated with understanding more about how the brain works to create our minds and thoughts. Not only our human brains, but also those of animals, including our dogs and cats. And it would be especially interesting to know precisely what makes one creature, or human, more intelligent than another.

It has long been recognized that the brain's neurons, or nerve cells, play a crucial role in the information processing that occurs *in the mind.* These are the specialized cells that transmit and receive electrical and chemical impulses from other neurons through the dendrites and axons[39] that interconnect the neurons. For a long time, it has been assumed that the greater the number of neurons, the greater the information processing capacity, and, therefore, the higher the cognitive abilities of the particular creature.

For decades, neuroscientists have conducted studies on a wide range of animals, including humans, to try to estimate the number of neurons in their brains, and therefore their relative intelligence. Not all have been that successful.

Initially, it was assumed that the structure of all brains was relatively uniform. Hence, the size of the brain of any creature was proportional to the number of neurons it had and, therefore, its cognitive ability. But this led to some problems. Elephants and whales have very large brains compared to humans. A sperm whale's brain, for example, is about 17 pounds, while a human brain is about three pounds. While whales are relatively intelligent creatures, their cognitive abilities are not thought to be even remotely

[39] **Dendrites** and **axons** are parts of neurons (nerve cells) in the brain and nervous system, and they play key roles in how the brain processes and transmits information. Together, dendrites and axons allow neurons to form complex networks that underlie thought, intelligence, consciousness, emotion, movement, and all brain functions.

comparable to those of humans. *So, pure brain size is not the answer.*

The next step was to correct for the size of the brain compared to the mass of the body it controlled. The idea was that the higher the *brain-to-body mass ratio* of an individual, the more brain capacity left over for cognitive functions.

To test the theory, neuroscientists measured the brain-to-body mass ratio in many animals. The results were generally in line with intuitive expectations. Whales, for example, had relatively high ratios, but they were lower than those of humans. Primates, such as chimpanzees, gorillas, and orangutans, were closer to the human ratio, but as expected, still lower. *One problem, however, was that small mammals, such as mice, have brain-to-body ratios similar to those of humans.* If their brains had a comparable structure, they would be much more intelligent than they appear.

The conclusion was that the brain-to-body mass ratio, and a later refinement of that measure, called *the Encephalization Quotient*, was a better representation of relative cognitive ability than just brain size, *but still imprecise for many of the creatures studied.*

It became clear that what was needed was a more effective method to measure the number of neurons in the brain directly. *Stereology*[40] is a standard tool in biosciences and has been used for many years to count the neurons in the brain, one small section at a time. Because the brain's neurons are not evenly distributed throughout its structure, *this method proved to be very slow and prone to significant errors.*

In about 2005, *a Brazilian neuroscientist made a breakthrough in our ability to count neurons quickly and accurately.* Dr. Suzanna Herculano-Houzel developed a method to strip out the nuclei of neurons from a brain sample

---

[40] **Stereology** estimates the number of neurons in the brain by analyzing systematically sampled, stained tissue sections under a microscope using the optical dissector method, which counts neurons within a defined 3D space. This technique provides unbiased and accurate results by applying strict counting rules and mathematical formulas to extrapolate totals from sample data.

using detergent. Then she homogenized the solution and used standard, electro-optical instruments to quickly and easily count the neurons. Officially, her method is called an *isotopic fractionator.* She prefers to call it by its more informal, yet accurate name, "brain soup."

Since then, she and her colleagues from Vanderbilt University have conducted a wide range of studies on comparative brain anatomy across various species, reaching some fascinating conclusions.

One of her major conclusions is that the human brain contains approximately 86 billion neurons, significantly more than any other species. Previously, neuroscientists estimated the number to be 100 billion, but no one had ever actually measured it before.

She also concluded that *the human brain is just a scaled-up primate brain,* not special in any other way. *Why did we evolve with this larger brain while other primates did not?* Brains, which are only about 2% of the body's weight in a human, require a substantial fraction, 25%, of the calories consumed by an individual. Because gorillas and orangutans have larger bodies than humans, and their low-calorie, vegetarian diets did not provide enough calories, even when they ate for eight hours per day, they *evolved* with a smaller brain, with proportionally fewer neurons, for their body size.

Humans, on the other hand, developed the habit of cooking food. This provided them with much *more densely packed calories* in the food they consumed, allowing them to support a much larger brain for their smaller body size. Evolution did the rest.

Dr. Herculano-Houzel's work also uncovered the reason why the body-to-mass ratio of mice, and for most rodents, is similar to that of humans. Yet, the creatures possess a level of intelligence lower than ours. *Rodent brain structure and scaling are the answers.*

While studying rodent brains, Dr. Herculano-Houzel found that *the* ***average size of a neuron*** *in a rodent brain increased with body size, so the brain size for larger rodents* *was larger than expected brains for the number of neurons it contained.*

In primates, like us, on the other hand, the neuron size remained the same as the body and brain size increased. *The result was that primate brains gained many more neurons than rodents with an increase in brain size.*

Examining primate brain structure in more detail, the cerebral cortex is the *outermost layer.* It is grey and has many folds. This is what you usually think of when you picture a brain.

It is this part of our brain that is responsible for our awareness, thought, intelligence, memory, and language. In other species, it is also the seat of higher thought processes. **To compare intelligence between species, *scientists agree we should focus on the number of neurons in the cerebral cortex not the whole brain*.**

Dr. Herculano-Houzel's findings in this regard are informative. *The cerebral cortex of a human brain contains about 16 billion neurons. Orangutans and gorillas have about 9 billion each, or 44% fewer. At the same time, African elephants are measured at 6 billion, or 63% less than the human population.*

So, which is more intelligent, your dog or your cat? It is tough to measure intelligence directly in creatures, but the number of neurons in the cerebral cortex is a good approximation. Dogs are estimated to have approximately 530 million cortical neurons, while cats have about 250 million, which is less than half that of a dog. Both creatures have only a small fraction of the neurons found in primates, about 4%, but dogs still outperform cats in comparison.

The human brain is a complex yet fascinating subject to study. The actual number of neurons contained in various parts of the brain is a critical component of our understanding. But it is not adequate to thoroughly explain intelligence and the differences in intelligence between humans. To gain better insights, extensive work is being done with new technologies to understand the details of how the brain actually processes and communicates information. But that is another story.

Update of the article

Over the past five years, Dr. Suzana Herculano-Houzel of Vanderbilt University has significantly advanced our understanding of comparative animal intelligence by applying her innovative neuron-counting method, *the isotopic fractionator*, to a diverse range of species. Her recent research continues to challenge traditional assumptions about brain size and intelligence, *emphasizing the importance of neuron count*, particularly in the cerebral cortex, as a more accurate predictor of cognitive abilities.

In a comprehensive study involving over 100 avian species, Dr. Herculano-Houzel and her collaborators quantified neuron numbers to explore the relationship between brain size and intelligence. They discovered that birds *with higher neuron densities in the pallial telencephalon*, a brain region analogous to the mammalian cerebral cortex, *exhibited greater cognitive capabilities.* This finding suggests that *neuron density, rather than sheer brain size, plays a crucial role in determining intelligence across species.*

Dr. Herculano-Houzel applied her neuron-counting techniques to fossil data *to estimate the cognitive abilities of extinct species, notably the Tyrannosaurus rex.* By comparing the estimated neuron counts in T. rex's telencephalon[41] to those of modern birds, such as ostriches and emus, she proposed that T. rex may have possessed neuron numbers comparable to those of primates, *such as baboons.* This *hypothesis challenges the traditional view of dinosaurs as unintelligent reptiles*, suggesting *a reevaluation of their behavioral complexity.*

---

[41] The **telencephalon** was a key part of the dinosaur brain responsible for sensory processing and motor control

In a study examining over 700 warm-blooded species, Dr. Herculano-Houzel found a *strong correlation between the number of cortical neurons and lifespan*. Species with higher cortical neuron counts tended to live longer and reach sexual maturity later, regardless of body size or metabolic rate. *This research suggests that neuron count is a more reliable predictor of longevity than traditional metrics, such as* body mass.

Dr. Herculano-Houzel continues to expand her research, applying her methodologies to various species to further elucidate the links between brain structure, neuron count, and cognitive function. Her work not only enhances our understanding of animal intelligence but also provides valuable insights into the evolution of the brain across different species.

Through her innovative approaches and interdisciplinary collaborations, Dr. Herculano-Houzel remains at the forefront of neuroscience research, challenging existing paradigms and offering new perspectives on the biological foundations of intelligence.

Exploring the dreamscape

Submitted to The Gloucester Daily Times but not yet published.

Everyone dreams while they are sleeping. In fact, it may be surprising to you that *we spend one and a half to two hours dreaming every night.* During these periods, the brain can generate a fully immersive, sensory-rich experience that feels as vivid and real as waking life, complete with sights, sounds, emotions, and even physical sensations. Dreams may play a role in our daily lives by helping us resolve emotional issues, engaging us in creative thinking, and contributing to the solution of some of our most pressing problems.

Each night, we experience four to six dream cycles over several hours, during which our brain enters the rapid eye movement (REM) stage of sleep. It is called *rapid eye movement* because, during this phase, your eyes move quickly under your closed eyelids. And *it is then that we have vivid dream experiences.* Each REM period gets longer as the night progresses. The first time may be only ten minutes. By morning, it increases to forty-five to sixty minutes. *That is a lot of dreaming, yet most of us remember extraordinarily little of these episodes.*

Most people's dreams are a mix of familiar and bizarre experiences. They may involve people we know, places we recognize, or be completely surreal scenarios. A common recurring dream for many people is realizing that there is a final exam tomorrow and knowing you have not attended classes or prepared for it. It is a great relief to wake up and realize that it was just a dream.

Another exciting part of the REM dream experience is called *lucid dreaming.* This is when a person becomes aware that they are dreaming while still in the dream state. In some cases, the dreamer can even control aspects of the dream, such as the environment, characters, or their own actions. Since REM sleep cycles become longer as the night progresses, lucid dreams are more likely to occur in the later stages of sleep, often in the early morning hours.

Most dreams are quickly forgotten as soon as we wake up. There are several reasons. Neuroscientists have studied the dreaming process in the brain and found that the chemicals that help form our memories are extremely low during dream times. This makes it harder for us to form memories of them. In addition, the brain's prefrontal cortex, which helps us process thoughts into memory storage, is less active during REM sleep. Furthermore, when we wake up, our brain quickly shifts focus from dream-content to real-world concerns, and this rapidly overwrites most memories of our dreams.

Scientists have concluded that this process of clearing out memories of dream content is helpful. It enables us to distinguish easily between real experiences and dreams. But even though we do not remember most of those images, dreaming plays a crucial role in our mental well-being.

Dreaming acts like overnight therapy for the brain, helping us process life's challenges and wake up feeling emotionally balanced and mentally refreshed. A dream may aid in our emotional processing by allowing the brain to work through complex feelings, unresolved conflicts, and stressful events. This can help reduce anxiety and improve our emotional resilience during waking hours. *Yet another benefit is that dreams allow our subconscious mind to work through our problems in unconventional ways. This is sometimes referred to as "sleeping on it."*

If we are ever sleep-deprived over a prolonged period and our natural course of dreaming is disrupted, scientists have found that we will experience significant negative effects on our mental health. The brain struggles to process emotions properly. This can lead to heightened anxiety, mood swings, and irritability. Our focus, decision-making, and creativity will become impaired, leading to difficulty concentrating and a fuzzy perception of critical issues. In extreme cases of prolonged sleep deprivation, people can experience hallucinations, paranoia, and symptoms similar to psychosis.

The discussion to this point has been about our dreams during REM sleep. However, distinct types of related sensory perceptions can occur during other parts of the sleep cycle. These come at either the beginning, when we are just falling asleep, called by scientists *the hypnagogic state*, or as we are awakening, *the hypnopompic state.* Only a small number of people experience these states, I have been one of them, but they are just as interesting and important as REM dreams.

As some people start to drift off to sleep, they may enter a hypnagogic state and experience the spontaneous appearance of visual or auditory sensations. *One of the significant differences of these events from REM dreams is that the observer knows they are in their bedroom and lying in bed.* They may see another person unexpectedly near their bedside, probably a stranger with unknown intentions. Or they may hear someone calling them.

Complicating the phenomena, some people undergo temporary sleep paralysis along with these experiences, making the event more intense or frightening. It can be very disconcerting to some people.

There is similarity between these experiences to the type of psychotic hallucinations associated with schizophrenia, although those are experienced in the waking state. For that reason, hypnagogic experiences, which are normal, are *no longer called hypnogogic hallucinations* but are referred to as experiences in the hypnagogic state.

To summarize, our nightly forays into the dreamscapes reveal a complex yet vital interplay between memory, emotion, and creativity. These ephemeral experiences, whether the vivid episodes of REM sleep or the fleeting images of hypnagogia, enable our minds to process daily stresses, heal emotional wounds, and spark innovative solutions, even if we only remember fragments upon awakening. Ultimately, understanding and valuing these nocturnal journeys underscores the indispensable role of sleep in nurturing our mental and emotional well-being, reminding us that even in our most unguarded moments, the brain is hard at work ensuring our resilience for the day ahead.

Update of the article

Using mindfulness to induce and utilize lucid dreaming –

Lucid dreaming, where individuals become aware they are dreaming and can sometimes control the dream narrative, has long fascinated scientists and the public alike. Recent research is shedding light on innovative techniques to induce lucid dreams, particularly *through the application of mindfulness during REM sleep.*

A 2024 pilot study led by researchers at Northwestern University and the University of Rochester Medical Center explored the potential of *reactivating mindfulness during REM sleep to induce lucid dreams.* The lead author was Dr. Ken Paller of Northwestern University.

Participants underwent a wake-back-to-bed protocol, engaging in a breath-counting task accompanied by ambient music cues. These cues were later replayed during REM sleep to reactivate the mindfulness context. The study found that this method *successfully induced signal-verified lucid dreams in two out of five participants,* suggesting that mindfulness-associated sensory stimulation during REM sleep holds promise for promoting lucid dreaming.

Further supporting the link between mindfulness and lucid dreaming, studies have shown that individuals with higher trait mindfulness[42] and meditation experience report increased frequency of lucid dreams. For instance, research indicates that long-term *meditators experience more frequent lucid dreams compared to non-meditators,* highlighting the role of sustained mindfulness practices in enhancing dream awareness.

The implications of these findings are significant. Lucid dreaming, facilitated by mindfulness techniques, may offer

---

[42] **Trait mindfulness** refers to an individual's inherent or dispositional tendency to be mindful in their daily life. Unlike state mindfulness, which is a temporary state of present-moment awareness often cultivated through practices like meditation, **trait mindfulness is a more stable and enduring personality characteristic.**

therapeutic benefits, including the treatment of nightmares, enhanced emotional regulation, and increased *creativity*. As research progresses, integrating mindfulness practices into sleep routines may become a viable method for individuals seeking to experience and utilize lucid dreams for personal growth and well-being.

In summary, the intersection of mindfulness and lucid dreaming represents a promising frontier in sleep research. By harnessing the power of mindfulness during REM sleep, individuals may unlock new dimensions of consciousness, offering both scientific insights and practical applications for mental health and creativity.

Exploring consciousness during the hypnagogic and hypnopompic states – The transitional phases between wakefulness and sleep, known as hypnagogic (as one falls asleep) and hypnopompic (as one awakens) states, are *characterized by unique sensory experiences and altered consciousness*. These fleeting moments often involve vivid imagery, auditory phenomena, and a blending of dream and reality, offering a *rich area of study for researchers interested in consciousness* and its applications in cognitive and therapeutic contexts.

Dr. Joseph Glicksohn of Bar-Ilan University in Israel has conducted extensive research into these transitional states. His studies indicate that individuals with *high levels of trait absorption*[43], a personality trait associated with deep engagement in sensory and imaginative experiences, are more likely to report occurrences of hypnagogic and hypnopompic phenomena. This suggests a link between specific personality characteristics and the frequency or intensity of these altered states of consciousness.

---

[43] **Trait absorption**, in psychology, refers to a personality characteristic marked by a *tendency to become fully engrossed in sensory and imaginative experiences*. Individuals high in trait absorption have a capacity for deep involvement in their mental imagery and are often easily captivated by their own thoughts and fantasies.

Complementing this psychological perspective, a collaborative study led by Dr. Christopher Timmermann and colleagues from institutions including King's College London and the University of Cambridge *proposes a neurophenomenological approach to understanding non-ordinary states of consciousness.* This framework emphasizes *the integration of subjective experiences with neural dynamics*[44], aiming to refine our comprehension of how these *transitional states can influence cognitive and emotional processes.* The researchers highlight the potential of such states to serve as catalysts for enhancing mental health interventions.

Historically, the significance of these transitional states has been recognized in various cultural contexts. Dr. David Hufford, professor emeritus at Penn State University College of Medicine, explored the "Old Hag" phenomenon[45], a cultural interpretation of sleep paralysis and associated hallucinations, in Newfoundland folklore. His work underscores the importance of considering cultural narratives when examining the subjective experiences associated with hypnagogic and hypnopompic states.

The exploration of these transitional states continues to offer valuable insights into the human mind. By bridging psychological traits, neural mechanisms, and cultural interpretations, researchers aim to deepen our understanding of consciousness and its potential applications in enhancing cognitive flexibility and therapeutic practices.

---

[44] **Neural dynamics**, refers to the *patterns of activity of neurons and neural circuits as they change over time.* Understanding these dynamic patterns is crucial for deciphering how the brain processes information and gives rise to our mental world.

[45] In Newfoundland, the **Old Hag** is depicted as a malevolent, witch-like figure, described as an old woman with a terrifying appearance, wrinkled skin, wild hair, and sometimes sharp teeth and glowing eyes. She visits people at night, specifically during the liminal state between wakefulness and sleep (i.e., during the hypnagogic or hypnopompic states).

Time travel is possible

First published by The Gloucester Daily Times on April 27, 2023

In 1895, the English author H.G. Wells wrote a science-fiction novella that became wildly popular. It was named *The Time Machine.* Wells set his story in then-present, Victorian England. In the tale, an adventurous inventor creates a machine that can bring him into the far future. There, he encounters a post-apocalyptic society populated by both beautiful people and underground monsters.

Ever since the work was published, the idea of being able to travel in time, both forward and backward, has captured the imaginations of people everywhere. Over the years, Wells' story has been made into two feature films of the same name, as well as two television adaptations.

But is time travel really possible? Not today, as described in Wells' tale. *But many scientists believe that the laws of nature allow it.* If only we could develop the technology to make it happen.

When we are children, we learn the basic facts of life. We realize that we live in the present, the here and now. The meanings of "here" and "now" become intuitive concepts. "Here" is the place we are in. "Now" is the present as defined by the time, day, and date on our smartphones.

Time can seem to pass slowly when we are in a boring situation. Or super-fast when we are with friends and having fun. We also know that time goes only in one direction, forward, and it can't be stopped.

But it turns out that the reality is much different from what we have come to believe. Albert Einstein, the world-famous physicist, published his special theory of relativity in 1905 when he was 26 years old and working as a patent clerk in Switzerland. Understanding the implications of relativity has profoundly changed how scientists perceive time. According to Einstein, time turns out to be an illusion. What!?

A key concept from the theory of relativity is that the three dimensions of space and one dimension of time are really part of the same thing. They are, in fact, fused together into the four-dimensional continuum that he called "Spacetime". *The implications of this concept are what allow us to consider the possibility of time travel.*

If you are sitting at your desk in your office on Monday morning at 9 am, that defines your location in Spacetime right now. But last Monday morning, you were also at your desk at 9 am. It is obvious that you were not at the same point in time, but neither were you at the same point in space. Remember that the Earth is traveling around the Sun while spinning on its axis, and the Sun is moving within the galaxy, the galaxy is moving relative to other galaxies. So, you and your office were at a very different point in absolute space at that particular time.

If you could trace your life in Spacetime, it would start at the point you were born and track through all the other points in Spacetime where you existed, till you reach the place and time of your death. Here is the mind-blowing part. According to Einstein, all of these points in Spacetime exist simultaneously. *The past, the present, and the future coexist simultaneously and are always present.*

He expressed this belief when, in 1955, he wrote a letter of condolence to the grieving family of his lifelong friend, Michele Besso. Einstein said, "*Now he has departed from this strange world a little ahead of me. That means nothing. For us, believing physicists, the distinction between past, present, and future only has the meaning of an illusion, though a persistent one.*"

Now, this idea completely violates your intuition about how the world works. However, based on numerous scientific experiments conducted over the last hundred years, the theory of relativity has been proven *correct in all its predictions.* So, it does indeed appear to represent reality.

Here is a way to think about this view of Spacetime that helps explain how time travel is possible. Consider your path along that Spacetime line from your birth to your death to be like a digital video recording. You can use your remote control to go into the video and pick any scene you want to play. Then consider you and the other people in that particular segment. They consider themselves to be in the present time. They remember what happened to them in past scenes but have no idea of what will occur later, even though the video is complete, and all those future scenes are there and will not change.

Since all points in Spacetime exist simultaneously, one can imagine that a way may be found to move from one point to another, either forward or backward. And time travel would be achieved. But how, exactly, can it be done?

The theory of relativity suggests a mechanism that could enable time travel. *It is called a "wormhole".* This is a structure that connects two different points in Spacetime and is based on a particular solution to the Einstein field equations.

To visualize how this would work, take a sheet of paper that represents a plane of Spacetime. Then draw two dots on one side of the paper, separated by several inches. The "wormhole" can be visualized by folding the paper (the plane of Spacetime) so that the two dots touch each other. You can see that it is now much easier to move from one point in Spacetime to the other because of the shortened distance.

Will we ever have a time machine that will allow us to visit other points in Einstein's "block universe" that represents all of Spacetime? History suggests that, given sufficient time, people will indeed develop the technology to achieve that goal. And if they do, they have probably been visiting us for millennia. In fact, some scientists, *such as Professor Michael Masters of Montana Technological University,* have speculated that time travelers, rather than aliens, may be one explanation for the Unidentified Anomalous Phenomena (UAP), also known as UFOs, which the Pentagon has reported as being real.

Update of the article

Scientific research on the possibility of time travel -

In 1988, a team of physicists at the California Institute of Technology, Kip Thorne, Michael Morris, and Ulvi Yurtsever, embarked on a *groundbreaking exploration into the theoretical mechanics of time travel.* Their goal was not to build a time machine, but to test whether the laws of physics, particularly those described by Einstein's general theory of relativity, would permit such a possibility. Their attention turned to the concept of a *traversable wormhole,* a hypothetical tunnel through spacetime that could connect two distant points, both in space and in time.

Unlike earlier versions of wormholes that were unstable or one-way, *Thorne and his colleagues sought to determine whether such a structure could be stabilized and rendered safe enough for a human traveler.*

The challenge was immense. According to their calculations, a *naturally occurring wormhole, if one existed, would collapse too quickly for anything to pass through it.* But there was a possible loophole: the introduction of *negative energy,* a strange concept rooted in quantum physics.

Negative energy[46] *is not just the absence of energy, but a state in which the energy density*[47] *is actually below that of the vacuum.* The *Casimir effect,*[48] A well-known quantum

---

[46] **Negative energy** - In quantum field theory, even "empty" space isn't truly empty. It's filled with fluctuating fields and virtual particles popping in and out of existence. *The energy of this vacuum is considered the baseline.* Negative energy, then, refers to a state where the energy is *lower than the vacuum,* something that seems paradoxical, but *can occur in very specific and extreme quantum conditions.*

[47] **Energy density** is a measure of the amount of energy stored in a given system or substance per unit volume or mass. It is a crucial concept in physics because it determines how much energy can be stored in a particular space or weight.

[48] The **Casimir effect** is a real and measurable phenomenon in quantum physics where two very close, uncharged metal plates placed in a vacuum *attract each other—without any external force applied.* It's caused by the

phenomenon observed in laboratory conditions *provided proof that such negative energy states can exist,* at least on very small scales.

Thorne's team proposed that if *enough negative energy could be generated and controlled, it might counteract the wormhole's natural instability and hold it open long enough for travelers to traverse.*

The implications of their work were profound. If one end of a stable wormhole were accelerated to near-light speed and then brought back, due to time dilation effects from Einstein's theory, a time difference would emerge between the two mouths. Entering the stationary mouth would then allow a traveler to exit at an earlier point in time through the accelerated one, essentially enabling time travel into the past. This scenario, while astonishing, *does not violate the known laws of physics on paper, which was a pivotal result.* It demonstrated that general relativity does not explicitly forbid time travel; it only demands extremely exotic and currently unattainable conditions to make it happen.

Though no such wormholes have ever been observed and the technology to generate or manipulate negative energy on a usable scale does not yet exist, the work of Thorne, Morris, and Yurtsever remains a cornerstone of modern theoretical physics.

It opened the door for serious academic inquiry into time travel, a subject once relegated to the realm of science fiction. Their research also helped refine the boundaries between what is physically impossible and what is simply technologically unfeasible, *for now.*

Other scientists who have expanded on the work of Thorne, et al. -

Many scientists have expanded on the work of Kip Thorne, Michael Morris, and Ulvi Yurtsever since their 1988 paper

---

strange behavior of *quantum vacuum fluctuations* and is one of the few laboratory demonstrations of **negative energy**.

on traversable wormholes and time travel. While no one has yet discovered practical ways to build a time machine, several physicists have explored the theoretical and quantum implications of their ideas.

Matt Visser of Victoria University, Wellington, New Zealand-

Visser is one of the most prominent researchers to follow Thorne's work. In the 1990s, he analyzed the structure and stability of traversable wormholes and suggested new geometries that might require less negative energy.

He demonstrated that "thin shell" wormholes, constructed by surgically grafting two spacetimes together, could potentially reduce the amount of exotic matter required. This helped make the concept of wormholes more mathematically tractable, although it remains highly speculative.

Sergei. V. Krasnikov of the P. N. Lebedev Physical Institute, Russian Academy of Sciences -

Krasnikov proposed a theoretical device called the Krasnikov tube,[49] which could potentially allow for faster-than-light travel. He built on Thorne's idea of modifying spacetime geometry to create paths allowing for backward time travel. His concepts still require negative energy or exotic matter, like Thorne's wormholes.

Stephen Hawking's Chronology Protection Conjecture (1992) - Hawking proposed that *the laws of physics may prevent time travel by destroying any time machine before it becomes operational (due to quantum effects).*

Hawking's proposal was a skeptical response to Thorne's work, which concluded that time travel isn't ruled out by general relativity. Thorne argued that, in General

---

[49] The paper was titled:
"Hyper fast Interstellar Travel in General Relativity"
Published in *Physical Review D*, Vol. 57, Issue 8, 1998

Relativity, specific spacetime geometries, like wormholes, rotating black holes, or the Krasnikov tube, could *allow for closed timelike curves* (CTCs).

Hawking argued that, on the other hand, quantum effects *might actually close the door. Hawking's conjecture* argues that when such a time-traveling geometry starts to form, *quantum effects become so extreme* that they destroy the time machine before it becomes usable.

The reason is that Quantum Field Theory in Curved Spacetime predicts that vacuum fluctuations (tiny bursts of energy from "nothing") near a CTC would grow rapidly. As the time machine becomes operational, *these fluctuations intensify exponentially, creating infinite energy densities at the boundary of the time loop.*

This intense energy would likely cause gravitational collapse or other catastrophic effects, preventing the loop from forming.

In simple terms, the universe sabotages any attempt to violate causality by creating conditions that destroy the time-travel structure.

Quantum Information Theorists (2000s–Present) -

Physicist Seth Lloyd of MIT and others have explored how *quantum mechanics* might handle the paradoxes that arise with time travel, such as the famous *grandfather paradox. This is where a traveler goes back in time and prevents their own existence.*

While general relativity allows for the possibility of time travel under certain conditions, *it doesn't address the logical contradictions it might create.* Lloyd's work *uses quantum theory to explore how those contradictions might be avoided.*

One idea involves the *Novikov self-consistency principle.* This idea suggests that *any actions taken by a time traveler were always part of history and therefore can't create a paradox.*

Another possibility is drawn from *quantum interpretations,* such as the *Many-Worlds* theory (described in the next chapter, "Quantum physics is 'spooky'"). The implication is that *traveling back in time could split the universe into different timelines, each consistent within itself, avoiding contradictions.* This idea appears to be consistent with the concept of the Multiverse.

Although these theories are highly speculative, they suggest that quantum physics may permit time travel without logical contradictions. In that sense, Lloyd's work helps reframe time travel as a question not just of engineering, but of how reality itself might behave at the quantum level.

Conclusion

While no one has turned Thorne's theory into a practical method of time travel, his framework has inspired decades of serious theoretical work. Scientists like Matt Visser refined the math of wormholes; others like Hawking have challenged the concept with quantum principles; and quantum theorists have investigated the paradoxes. The core problem remains: *stabilizing a wormhole or time machine still seems to require exotic conditions or negative energy, which we cannot currently produce at scale.*

Quantum physics is "spooky"

First published by the Gloucester Daily Times on March 7, 2020

The idea of parallel universes has been a staple of science fiction for many years. Movies, books, and TV programs, such as "Fringe" and "Counterpart," have explored this theme. In these storylines, the key characters meet their "selves" from the other world.

You may not realize it, but there is some scientific evidence to support this unusual idea. It originates from the *Many-Worlds Interpretation of quantum physics*, which is gaining support among many physicists. More about this later.

What physicists do for a living is to observe what is going on in nature and develop predictive, mathematical descriptions of that behavior. They can then show how that behavior will change as conditions change.

In the early twentieth century, quantum physics evolved using this approach. It happened when physicists found that classical physics couldn't explain what they saw when they were working with tiny particles like electrons.

The mathematical descriptions that physicists developed to characterize electrons' actions were very different from those of classical physics. The result was that the *more we explore the implications of quantum physics in our reality, the weirder and more mysterious it seems.*

You may not recognize it, but quantum physics has had a profound impact on our daily lives. Many of the advanced technology products we use today, like computer chips and lasers, have resulted from our understanding of quantum physics. We are now on the verge of dramatically increasing our computer power with a revolutionary approach that relies even more directly on quantum concepts.

This new approach, known as quantum computing, is significantly different from classical computing. Our current laptops and iPhones have chips that encode their programs using "bits". These bits can be made to be either a zero or a one. Programmers use this capability to generate large, sophisticated programs that solve complex problems.

However, we are now reaching the limit of this technology's capabilities to perform those complex calculations. So, computer scientists at companies like IBM, Microsoft, and Google are turning to the use of *quantum bits*, or *qubits*, to make a *major breakthrough in computer power*.

These qubits are quantum systems that utilize some of the spooky properties of quantum physics, including "superposition"[50] and "entanglement"[51], to encode the programs. This approach will enable a quantum computer *to conduct a massive number of calculations concurrently.* Scientists will eventually be able to *solve* problems in areas such as medicine, chemistry, and weather that are currently impossible with today's computers, including the largest supercomputers. *And the result will be significant discoveries in many fields.*

One of the quantum properties used in qubits is "superposition". Instead of just a one or a zero, a qubit can be in multiple states at the same time. It's hard to believe, *but a qubit can be both a one and a zero simultaneously!* An analogy to the idea of quantum superposition is when you pluck two strings of a guitar simultaneously. The sound waves you hear are the combination of the two sounds, or what is termed a superposition of the two notes.

How is the phenomenon of superposition even possible in the qubit? It is precisely because *quantum systems, like electrons, behave like waves,* similar to the sound wave in a superposition of the two notes coming from the guitar. *In*

---

[50] **Quantum superposition** is a fundamental principle of quantum mechanics stating that a quantum system—like an electron, photon, or atom—can exist in multiple possible states at the same time until it is measured or observed.

[51] **Quantum entanglement** is a phenomenon in quantum mechanics where two or more particles become linked in such a way that the state of one particle instantly influences the state of the other—*no matter how far apart they are.*

fact, an electron's normal state is a wave in superposition of all its possible properties.

Electrons behave as waves, that is, *until physicists* ***observe*** *them during an experiment* !? *Then the electrons suddenly change and behave like particles. Yes, that's right.* An electron is a wave, except when a human observes it. The electron appears to be aware that it is being observed, and its wave function, which describes its normal behavior, collapses; the electron then acts like a particle. ***Now, that is bizarre.*** This approach to quantum physics is called the Copenhagen Interpretation.

Perhaps even more strange is the quantum phenomenon of entanglement. *Two quantum systems, like those found in a qubit, can interact and become entangled.* That is, they are inextricably linked in perfect unison. A change in one quantum system results in an instantaneous, related change in the other. That is hard to explain with classical physics. But even more mystifying is that they *continue to be linked in this way even if the two quantum systems are as far apart as opposite ends of the universe*!

This observation really bothered one of the most famous physicists of all time, Albert Einstein, creator of the Theory of Relativity. Einstein is also one of the founders of quantum physics. In fact, his only Nobel Prize was for quantum physics, not relativity. *What bothered Einstein was a conflict between the findings of these two fields.*

One of the significant results of his relativity work was to show that the speed limit for anything in the universe is the speed of light. This is approximately 186 thousand miles per second. Nothing can exceed that number. Knowing this fact, he found it hard to accept that two quantum systems, separated by the diameter of the universe, could instantaneously know that their counterpart have changed. At the speed of light, it would take 93 billion years for the two entangled particles to communicate with each other! As a result, *Einstein termed entanglement to be "spukhaften fernwirkung" or "spooky action at a distance".*

Perhaps the spookiest revelation of quantum physics is one that has been gaining increasing acceptance among many physicists. It is called the *Many Worlds Interpretation*.

In 1957, a young graduate student named Hugh Everett objected to the Copenhagen Interpretation in his doctoral thesis. Everett felt that *the idea of the quantum system's wave collapsing* was contrived. His alternative explanation was very simple and straightforward relative to the mathematics. He assumed that the whole universe, and everything in it, including us, is a wave in superposition, and that the wave function is real and doesn't collapse.

This interpretation implies that all possible outcomes of any event are in superposition within the universal wavefunction. And, that all outcomes are physically realized, including us in them, in some "world" or universe.

To the observer in each universe, it just seems that the wave function collapsed because they are not aware of the other worlds. It is truly astonishing to think that there is more than just the universe we know, *but that is the implication of the mathematics*!

Since there are so many quantum events occurring all of the time, the number of universes is innumerable. And there is a "you", as an observer, in most of them. While challenging to accept, that is what the straightforward mathematics in this interpretation tells us. Remember that, once upon a time, people thought that the sun revolved around the Earth because that is what their everyday experience told them.

The concept of many parallel worlds has been explored by numerous science fiction writers, who have examined the theme of meeting the person who is "yourself" from another world. In one sense, the Many-Worlds Interpretation of quantum physics supports this science fiction view since it provides a scientific rationale for parallel universes. But scientific formulation also does not allow the worlds to interact, so rest assured, in real life, you will never meet any of your other selves.

Update of the article

Several prominent physicists and philosophers of science support the Many-Worlds Interpretation (MWI) of quantum mechanics, originally proposed by Hugh Everett III in 1957. These researchers have produced peer-reviewed articles, books, and conference presentations that develop, defend, or explore the implications of the Many-Worlds Interpretation (MWI). Here are some of the key figures and what they're saying:

Bryce DeWitt, University of Texas at Austin (*deceased 2004*)-

Bryce DeWitt was a pioneering physicist who played a crucial role in developing and popularizing the Many-Worlds Interpretation (MWI) of quantum mechanics following its original proposal by Hugh Everett III. While Everett introduced the core mathematical and conceptual framework in his 1957 doctoral thesis, it was DeWitt who gave the interpretation its now-famous name and helped bring it into the scientific mainstream during the 1970s and 1980s.

DeWitt strongly agreed with Everett and believed that the apparent "collapse" of the wave function in quantum mechanics was an illusion, and that quantum theory should be taken literally: all possible outcomes of a quantum event truly occur, each in a distinct, branching universe. *He was one of the first physicists to publicly and persistently advocate for MWI as a serious scientific theory rather than a philosophical curiosity.*

DeWitt's most influential work on the subject was his edited volume, **The Many-Worlds Interpretation of Quantum Mechanics** (1973), co-edited with Neill Graham. This book compiled key foundational papers on MWI, including Everett's original thesis, commentaries by other physicists, and interpretive essays by DeWitt himself. It served as both a historical archive and a comprehensive defense of the interpretation, making Everett's ideas more accessible to a new generation of physicists and philosophers. DeWitt's vivid imagery, particularly the metaphor of the universe "constantly splitting" into countless realities, helped capture the public and academic imagination. While critics often saw his language as overly speculative, DeWitt saw it as a necessary reflection of what quantum theory actually demands when taken seriously. His work laid the groundwork for the modern revival of MWI, influencing contemporary thinkers like David Deutsch, Sean Carroll, and David Wallace.

David Deutsch, University of Oxford, pioneer in quantum computing -

Deutsch is one of MWI's most vocal advocates, arguing that *quantum computing* only makes sense if MWI is true, *because quantum algorithms explore many paths at once in a multiverse.*

His book, **The Fabric of Reality** (1997), is a groundbreaking and provocative work that attempts to unify our understanding of the universe. Rather than presenting science as a collection of disconnected facts or theories, Deutsch proposes that there are *four central strands, four "deep" theories,* that, when understood together, offer a more profound and accurate picture of reality than any one of them alone. These are: the *Many-Worlds Interpretation (MWI)* of quantum mechanics, *Popperian epistemology,*[52] *Turing-based computation* and *Darwinian evolution.*

His bold claim is that these ideas, if taken seriously and literally, don't just describe reality — ***they are the fabric of reality***.

One of the book's most significant contributions is its passionate and detailed defense of the *Many-Worlds Interpretation.* Deutsch argues that MWI is not just one interpretation among many, but *the only one that actually*

---

[52] **Popperian epistemology** refers to the theory of knowledge developed by philosopher *Karl Popper*, which is primarily centered on the idea of *falsifiability* as the key criterion for scientific knowledge. It stands in contrast to traditional inductivist views of science.

takes the mathematics of quantum theory at face value. He explains how *quantum interference, computation, and the behavior of particles can only be meaningfully understood if we accept that all possible outcomes of quantum events actually occur — in parallel, branching universes.*

By developing this argument through vivid metaphors and logical clarity, Deutsch helped bring MWI from a fringe idea into more serious philosophical and scientific consideration, inspiring physicists like *Sean Carroll and David Wallace* to carry the torch further.

Sean Carroll, Johns Hopkins University; formerly at Caltech

Carroll argues that MWI is the simplest and most elegant explanation of quantum mechanics because it doesn't require wavefunction collapse or ad hoc additions.

He has written the popular science book *"Something Deeply Hidden," which delves into his support for MWI in detail.* In it, Carroll argues that MWI is the best way to understand what quantum mechanics is *actually telling us about reality,* without resorting to vague notions like "wavefunction collapse" or artificially dividing the world into quantum and classical domains.

For Carroll, the elegance of quantum theory lies in its *simplicity and determinism,* and the Many-Worlds framework respects both, while providing a consistent and literal interpretation of the mathematics.

Carroll sees the *universal wavefunction* as the single, real physical entity, evolving smoothly and deterministically according to the Schrödinger equation. When a quantum event happens, such as a measurement, the wavefunction doesn't collapse. Instead, ***the universe*** "branches," with each possible outcome of the event *realized in a separate, non-interacting world.* According to Carroll, *this branching isn't metaphysical fantasy; it is what naturally follows if you take the quantum formalism seriously. The appearance of randomness or chance is simply the result of an observer becoming correlated with one of many possible outcomes,* with their "copies" in other branches experiencing the other outcomes.

One of Carroll's most important points is that *MWI doesn't add anything* to quantum theory, in contrast to interpretations like Copenhagen (which introduces a special role for observers) or Bohmian mechanics (which adds hidden variables). In Carroll's view, other interpretations are "epicycles"[53], complicated fixes to *avoid accepting the weird but mathematically simple consequences of quantum mechanics*. He argues that physicists have been reluctant to embrace MWI not because it fails scientifically, but because *its implications about reality, such as the existence of countless parallel versions of ourselves, are unsettling*. But for Carroll, discomfort is not a good reason to reject a theory that is otherwise logically coherent and empirically indistinguishable from its competitors.

Ultimately, Carroll believes that MWI is not only true to the spirit of quantum theory but also opens the door to more profound insights into time, entropy, and the structure of reality itself.

David Wallace, University of Pittsburgh, formerly the University of Oxford -

David Wallace views the Many-Worlds Interpretation as the most coherent and natural interpretation of quantum mechanics, one that takes the theory at face value without introducing arbitrary mechanisms, such as wavefunction collapse. In his view, the mathematical formalism of quantum theory, particularly the Schrödinger equation, describes a single, unified wavefunction that evolves deterministically. Rather than interpret the probabilistic outcomes of measurement as fundamental randomness, Wallace argues that the universe *actually branches* into

---

[53] **Epicycle** is often used metaphorically to describe a complicated workaround added to a theory to preserve it in the face of contradictory evidence, rather than revising the underlying assumptions.

multiple, non-interacting versions during quantum events. Each of these branches represents a different "world" in which a particular outcome is realized.

These worlds, however, are not fundamental building blocks of reality but rather *emergent structures.* Drawing an analogy with how classical objects like waves or hurricanes arise from more basic physics, Wallace suggests *that "worlds" are patterns that emerge within the universal wavefunction through the process of decoherence.*[54] Decoherence explains how quantum superpositions effectively become separate, stable realities, each with their own version of observers, measuring devices, and histories. *In this way, MWI avoids the need for a special "collapse" mechanism while still accounting for the classical world we experience.*

One of Wallace's most important contributions is his defense of how *probability* works in this deterministic multiverse. If all outcomes happen, what does it mean to say one is "likely"? Wallace answers this using *decision theory*, showing that a *rational agent* in a branching universe should act as if the standard quantum probabilities (given by the Born rule[55]) apply. Even though all branches exist, *an observer doesn't know in advance which branch they'll find themselves in*, a situation called *self-locating uncertainty*.

Ultimately, Wallace argues that the Many-Worlds Interpretation offers a cleaner and more parsimonious understanding of quantum mechanics than its rivals. It remains faithful to the core mathematical structure of quantum theory, accounts for measurement without resorting to mystery, and explains our experiences without requiring ad hoc additions, such as wave collapse. While it challenges our everyday intuitions, particularly with its

---

[54] **Decoherence** in quantum theory is the process by which a quantum system transitions from behaving "quantum-like" to "classical-like." It plays a crucial role in explaining why we don't observe quantum superpositions in our everyday world.

[55] The **Born rule** is a fundamental principle in quantum mechanics that tells us *how to extract probabilities from the wavefunction* of a quantum system.

vision of countless parallel worlds, Wallace believes this is a small price to pay for a theory that is logically consistent, empirically adequate, and philosophically robust.

Lev Vaidman, Tel Aviv University -

Vaidman is one of the most active and rigorous defenders of the Many-Worlds Interpretation of quantum mechanics. He has made significant contributions to the philosophical, mathematical, and physical foundations of quantum theory, particularly in defense of the Many-Worlds Interpretation (MWI) as the *most natural, consistent, and deterministic interpretation of quantum mechanics.*

He has numerous peer-reviewed papers, including in the journals *Foundations of Physics, Studies in History,* and *Philosophy of Modern Physics.* One example is the article, *"Many-Worlds Interpretation of Quantum Mechanics"* (Stanford Encyclopedia of Philosophy, a scholarly, but not peer-reviewed, journal)

Vaidman frequently defends MWI as the most coherent interpretation for the field of *cosmology.*[56]

[56] **Cosmology** is the scientific study of the *origin, structure, evolution, and ultimate fate of the universe.* It seeks to understand the universe *on its largest scales*—galaxies, clusters, the cosmic web—and how it all began, how it's changing, and where it's headed.

Conclusion to Part V

As we close this section of *The Curious Mind*, we are left standing at a remarkable intersection, where cutting-edge science meets the enduring mysteries of existence. From the blood-stained linen of the Shroud of Turin to the shimmering web of quantum entanglement, from the coded patterns of regenerative medicine to the blurred boundaries between dream and waking, we find ourselves confronted not only with new data but with profound questions: What does it mean to be alive? What does it mean to be conscious? And are we, perhaps, brushing up against the edges of something larger, something that science alone cannot fully measure, yet cannot ignore?

In the chapters on Intelligent Design and the Shroud, we dared to ask whether nature bears the fingerprint of a Creator, not as a matter of faith alone, but as a question open to scrutiny. In medicine, we saw how our growing command of biology and technology may soon allow us to repair and perhaps even indefinitely sustain the human body. The once-impossible notion of "escape velocity" for aging now feels less like science fiction and more like a future we must ethically prepare for. And in a deeply personal turn, "The Day I Flatlined" reminded us that the boundary between life and death, once seen as absolute, is now more fluid, more technical, and perhaps more mysterious than ever.

Meanwhile, physics has shown us that reality may be stranger than we imagined. If the laws of physics do not forbid time travel, what does that mean for causality, memory, and fate? If quantum particles can affect each other instantaneously across space, how should we understand the fabric of reality itself? In exploring these puzzles, we are not merely seeking clever answers; we are pursuing a more profound coherence between the universe and our place within it.

What emerges from these chapters is not a contradiction between science and meaning, but a *conversation,* one that acknowledges the limitations of reductionism[57] while celebrating the power of reason. The deeper we probe, the more it seems that nature is not only comprehensible but astonishingly rich in structure, possibility, and perhaps even intention. Whether or not one sees the divine in that complexity, there is awe to be found, awe that demands both intellectual humility and fearless curiosity. *The Curious Mind* thrives in that space where mystery invites exploration, and exploration, in turn, deepens the mystery.

[57] **Reductionism** is a philosophical and scientific approach that involves explaining complex phenomena by breaking them down into their simpler, more basic components.

Part VI

Science, War, and Invention: Lives that Changed the World

VI. Science, War, and Invention: Lives that Changed the World

Introduction to Part VI

True breakthroughs are often born at the intersection of genius and courage in the history of science and innovation. The lives of great thinkers, inventors, and leaders are marked by moments when they chose to challenge the status quo, confront impossible odds, or pursue ideas that others deemed impossible. In this section, we explore the stories of such individuals, men whose brilliance changed their fields and left a profound impact on history itself.

We begin with Yuri Tuvim, a man whose journey from the Soviet Union to becoming an American patriot reveals how personal transformation can spark a deep sense of purpose. His life is a testament to the power of love for one's adopted country to inspire remarkable achievements.

Alongside Tuvim's tale, we uncover the little-known but critical role of British citizens, who later lived on the North Shore of Boston, in helping to decode the Enigma machine, a feat of cryptographic brilliance that shortened World War II and saved countless lives. The group included Dr. Gordon Welchman of Newburyport, who, with Alan Turing, played a key role in breaking the Enigma Machine.

From the Boston and Annisquam laboratories of Dr. Isaac Adams, the inventor of nickel plating, to the electrifying innovations of Alexander Graham Bell, who lived in Salem on the North Shore while inventing the telephone, we delve into stories of creative ingenuity that redefined technology.

Yet the same brilliance that sparks invention can also raise profound ethical questions. J. Robert Oppenheimer, the mind behind the atomic bomb, stands as a complex figure in this collection; his story reveals the dual-edged nature of scientific power.

Finally, the narrative turns to the lives of the daring "Jumping General", James Gavin, commander of the 82nd

Airborne during World War II, creator of the U.S. Army's manual for Parachute Tactics: and John Hays Hammond of Gloucester, the brilliant inventor who revolutionized remote control technology that is so prevalent in today's unmanned drones. Each of these figures, whether on the battlefield or in the workshop, exemplifies the courage to dream, the resilience to struggle, and the brilliance to reshape the world. Together, these stories reveal not just the minds of great men but the spirit of discovery and responsibility that defines true greatness.

Who invented the electric light bulb?

First published in the Gloucester Daily Times on January 1, 2016

If you ask any schoolboy or schoolgirl, "Who invented the electric light bulb?" they would say, "Thomas Edison." Edison, perhaps the most famous and successful inventor of modern times, was asked late in his career to settle a debate over what he considered his greatest invention. He replied without hesitation, "Incandescent electric lighting and power system." Some scholars, however, believe that Dr. Isaac Adams Jr., a long-time resident of Gloucester, had invented a working, electric light bulb in 1865, fourteen years before Edison produced his first successful bulb in 1879. With a minor exception, the design of Adam's incandescent light bulb was virtually the same as Edison's invention. And yet, very few people have ever heard of Dr. Adams' success.

Dr. Adams was born in Boston in 1836. His father, Isaac Adams Sr., was already a well-known inventor and businessman. The elder Adams played a significant role in his son's also becoming a famous inventor and entrepreneur, although he did it unwittingly.

The elder Adams was a powerful, larger-than-life man who owned several companies in Boston. His primary enterprise was I&S Adams Company, a manufacturer of printing presses. One of his "Acorn" manual presses can still be seen at the Cape Ann Museum.

As a relatively young man, Isaac Sr. invented and had patented the *first powered printing press for fine book printing*. It was a breakthrough that became highly successful because it led to the production of relatively inexpensive books, making literature accessible to everyone. As a result, Isaac Sr. became famous and one of the wealthiest men in Boston.

Initially, the strong-willed father *did not want* his son to follow in his footsteps. Instead, he wanted the young man to be a medical doctor and a gentleman. It was Isaac Sr.'s early life that led him to make such a plan for his boy.

Isaac Sr. was born into a large, poor farm family in rural New Hampshire. After his father died, the youngster had to help support his mother and siblings. At the age of 9, he left home to work in a cloth-making factory in Rochester, New Hampshire. As a result of his impoverished beginnings, Isaac Sr. grew up wanting to become wealthy and to have a son who would be the highly educated and refined "gentleman" that he could never be.

This is what drove the now successful elder Adams to put his son through a rigorous schooling experience. After attending private preparatory schools in Sandwich, New Hampshire, and Boston, young Adams graduated from Bowdoin College and, at his father's behest, went on to Harvard Medical School. While he was at medical school, young Isaac met Professor Josiah Cooke, who changed his life and set him on the path to becoming an inventor. Cooke sparked the young man's interest in the field of electrochemistry. With Cooke's encouragement, Isaac Jr. worked on the problem of plating base metals with nickel. *While he was still in medical school, Isaac Jr. developed the first workable nickel-plating process.* It was an invention for which he would later receive a highly valuable U.S. patent, which would make his fortune.

How Isaac Jr. also came to invent the first incandescent bulb had its roots in the next phase of Dr. Adams' education. The new medical doctor graduated from Harvard just as the Civil War was beginning. His father then sent him to another medical school, this time in Paris. Isaac Sr. wanted to get his son away from the war and increase the young man's medical knowledge. Isaac Jr. did as his father wished. But in reality, he didn't want to be a physician. Instead, he wanted to pursue a career in technology. The young man's success with nickel plating, his strong educational background, and the deep admiration he had for his father, an inventor and entrepreneur, led Isaac Adams, Jr., to consider such a goal. At this point, however, he didn't mention his intentions to the elder Adams.

In Paris, Isaac Jr. studied advanced medical techniques, but he also became familiar with a technology that would become the basis of his electric light bulb invention. During his studies, young Isaac met Monsieur Adolphe Gaiffe, a Paris-based instrument maker and chemist. What intrigued Isaac the most among Gaiffe's technological bag of tricks were Geissler tubes. These are gas-filled glass tubes with electrodes sealed at each end. When a current is passed through the tube, the gas becomes "excited" and emits a low-level illumination, like a neon light. These devices were used at the time to aid in illuminating surgeries.

The young doctor became proficient in glass blowing and constructing the Geissler tubes. When he returned to Boston, he briefly established a medical practice to please his father, but he soon turned to technology to support his livelihood. He designed and sold a range of Geissler tubes to several Boston companies for use in producing light for various applications. This led young Isaac to embark on a series of experiments aimed at developing an incandescent bulb.

He produced his first successful lamp using a carbon filament in 1865. However, he then abandoned the technology, didn't seek a patent, and went on to develop other, more lucrative products that exploited his patent on nickel plating.

So, why isn't Dr. Isaac Adams Jr. recognized for his contribution to the invention that Edison considered to be his greatest accomplishment? Perhaps part of the answer lies in the character and integrity of Dr. Adams.

In 1890, Dr. Adams had retired from the nickel-plating business and was living in Annisquam. He was contacted by the law firm representing the United States Electric Lighting Company in their suit against Edison over the invention of the electric light bulb. The lawyers hoped that Dr. Adams would claim that he had invented the carbon filament electric light bulb before Edison. If the claim were to be upheld, it would have denied Edison the patent and his monopoly on the production of light bulbs.

Unfortunately, for the lawyers, Dr. Adams's penchant for truth, honor, and humility led him to describe his work during the patent trial in only the most accurate terms, when he could have easily made a case for his developments anticipating the Edison invention.

He testified that his intention in researching the bulb was to develop glass sealing technology, not necessarily to produce an electric light bulb. And, besides, he said, he had no means to generate the power to provide a viable electric lighting system. So, that is why he never patented his bulb.

The result of Dr. Adams' testimony was that the court found Thomas Edison to be the first to develop, patent, and exploit the technologies for the electric light bulb and power system. And Edison was awarded patent rights. So, most people then just forgot about Dr. Adams' contribution to electric lighting.

Dr. Adams passed away on July 24, 1911, in Annisquam. He is well remembered as the inventor of the nickel-plating process. However, you now know that he also made other significant developments, including a working electric light bulb fourteen years before Thomas Edison.

Update of the article

The later years of Isaac Adams, Sr. - Late in his life, Isaac Adams Sr., renowned for inventing the Adams Platen Press, retired to Center Sandwich, New Hampshire, where he had apprenticed as a very young woodworker. After a successful career in Boston as an inventor and manufacturer, he returned to his roots in the Lakes Region. There, he acquired extensive farmland consisting of 2,600 acres and established a substantial estate, reflecting both his wealth and his enduring connection to the area.

Adams's retirement was not a withdrawal from public life. He had previously served in the Massachusetts Senate in 1840 and remained engaged in civic matters. His involvement with the American Emigrant Aid Company, which supported anti-slavery settlers in Kansas, indicates his continued interest in national issues.

Isaac Sr. had an active interest in politics, as evidenced by his election to both local and state offices. He was elected a Selectman in the Town of Center Sandwich and also became its Town Meeting Moderator. The people in the area also elected him as a State Representative. This meant he was required to be in Concord, the capital of New Hampshire, when the General Court was in session.

When at home, most of his time was spent managing the numerous projects on his vast estate, including building an enormous granite wall that was wide enough for a team of oxen to walk on top of. Then, he made Citizen Kane-like trips to Europe, where he purchased marble statues that he placed as decorations on the walls. One of those, a statue of the mythical Niobe,[58] has been restored and can still be seen on the wall.

Adams passed away in Sandwich on July 19, 1883, at the age of 80, with his medical doctor son by his side. His legacy includes not only his contributions to printing technology but also his influence on the local community. The Sandwich Historical Society has documented many aspects of his life and estate, preserving the memory of a man who significantly impacted both industry and his hometown.

Dr. Adams' extended career in science and technology - Dr. Isaac Adams, Jr., enjoyed an enduring career in science and technology. Beyond his groundbreaking work in nickel electroplating, Dr. Adams was a prolific inventor whose innovations spanned multiple scientific and industrial domains. Trained as a physician, Adams ultimately channeled his talents into electrochemistry, mechanical engineering, and applied physics, leaving a legacy of diverse technological contributions.

In the mid-1860s, Adams began experimenting with Geissler tubes, early gas-discharge devices used in scientific demonstrations and research. He manufactured these tubes commercially between 1865 and 1868, leveraging skills he acquired while studying in Paris under the instrument

---

[58] **Niobe** is a tragic figure from Greek mythology, known for her story of pride, loss, and transformation. Grief-stricken, Niobe was transformed by the gods into a stone figure. Even in this stony form, she continued to weep eternally, her tears becoming a symbol of her unending sorrow.

maker Adolphe Gaiffe. This led to his invention of the carbon filament electric light bulb fourteen years before Edison.

Adams also made significant strides in firearms technology. His work on breech-loading rifles was considered nearly as important as his electroplating innovations. He secured a patent in 1869 for improvements in the manufacture of metallic firearm parts, reflecting his broader interest in metallurgy and mechanical systems.

His inventive output extended into rubber as well. In 1879, he patented a method for coating metallic articles with vulcanizable rubber. This patent illustrates his persistent engagement with emerging technologies of the 19th century.

Other diverse creative projects included the invention of glass bricks for use in construction and participation in a worldwide, sophisticated measurement of the Earth's dynamic magnetic field.[59]

Using sand from local beaches, Adams experimented with glass to replace clay in bricks. The resulting product was translucent, allowing some light to enter the house.

The British government, a maritime nation, undertook a long-term, worldwide project to obtain data and understand how the Earth's magnetic field was changing. This was important for improving navigation. Dr. Adams designed and built a sophisticated magnetometer to make measurements. Then he constructed a special observatory near his home in Annisquam, Massachusetts, where he collected data over many years and shared it with the international project.

Throughout his career, Adams exemplified a multidisciplinary approach to invention, seamlessly integrating his medical education with his expertise in mechanics and chemistry. His wide-ranging innovations underscore a legacy that extends well beyond nickel plating,

---

[59] See the chapter, "The Earth's magnetic field is on the move" in Part IV of this book.

marking him as a versatile and influential figure in the annals of American industrial history.

Alexander Graham Bell on the North Shore

First published in the Gloucester Daily Times on August 17, 2018

Many cities want to claim some part of Alexander Graham Bell's distinction as the inventor of one of the most important technological developments of all time, the telephone. Three of those locations are Boston and Salem in Massachusetts and Brantford, Ontario, in Canada. They each have strong rationales to consider themselves the birthplace of the telephone.

Alexander Graham Bell's journey to his monumental invention began when he took on 5-year-old George Sanders of Salem as a pupil on October 1st, 1872. The boy was born totally deaf and had never spoken a word in his life.

Alec Bell, as he was called, came from a family steeped in developing methods to teach the deaf to communicate. So, the twenty-five-year-old Bell was well qualified for the task. His pupil was the son of Thomas Sanders, a wealthy leather merchant who was to become a key financial investor in Bell's telephone technology. Eventually, Alec moved from Boston, where he was teaching, into a room at Sander's mother's house, at 242 Essex Street in Salem, so he could be near the boy.

Bell had always been interested in technology. He learned that the telegraph industry was desperate to have the capability to transmit multiple messages over a single wire. He investigated the field and began conducting experiments at night in Sanders' house after finishing his day's work tutoring young George. These developments would ultimately lead to the design of the telephone, allowing Salem to claim a key role in its birth.

While looking to buy apparatus for his experiments, he heard about a laboratory in Boston owned by Charles Williams. On a visit to Williams' operation at 109 Court Street, he met Thomas A. Watson, a Salem native and resident. The young man was an accomplished machinist and electrical worker and could be a great help to Bell. Being very impressed with

Watson, Bell invited the young man to collaborate with him on the multiple message telegraph project. Thus began the most important collaboration of Bell's life.

In the summer of 1874, Alec was visiting with his parents in Brantford, Ontario. It was then that he decided to drop his work on the telegraphy development and turn his attention to transmitting voice over a wire.

He combined all that he had learned from the telegraph experiments and about the physics of sound from teaching the deaf, and came up with the basic idea of the telephone. *Many years later, in a speech given in **1916**, Bell recalled that moment as the birth of the telephone.* And so, Brantford feels justified in claiming the title of Telephone City.

But it was to take a considerable amount of work over the next two years in Bell's Boston laboratory to get the device to work. The lab was in an attic space above Charles Williams' shop on Court Street.

As Bell and Watson began to close in on a working model of the telephone, they became concerned about the security of their design. So, they moved to a more private location at 5 Exeter Place, about one-half mile from the Williams operation. *It was there on March 10, **1876**, when Bell spoke his famous first sentence ever heard over a telephone.* According to Mr. Bell's journal in the Library of Congress, the conversation was *"Mr. Watson, come here, I want to see you." So, Boston holds a preeminent claim as the literal birthplace of the telephone.*

Salem played yet another role in the development of telephone technology. On February 12, 1877, almost one year after the first sentence was spoken over the device, Bell did the *first public demonstration of his invention.* The site was the *Salem Lyceum,* now Turner's restaurant on Church Street. He called Watson over a telegraph line to the offices of the Boston Globe, about 35 miles away. Watson spoke to the Lyceum audience over the telephone and even sang to them.

Later that night, a Boston Globe reporter used the same telephone in Salem to call in a story to the paper in Boston.

That was the first instance where the phone was used for such a purpose.

The Salem Lyceum demonstration of a long-distance call, however, *was not the first.* On August 10, 1876, six months earlier, Bell made a *private test call* from his parents' home in Brantford, Ontario, to Mr. Watson in a neighboring town, Paris, Ontario. It was about eight miles away, and the test used an existing telegraph line.

Bell's fame and wealth grew over the next few years as he worked with his investors to secure his patent and to establish his telephone company. Now that he was earning a substantial living, on July 11, 1877, he married his long-time sweetheart and pupil, Mabel Hubbard. She was the daughter of one of his investors, Gardner Hubbard, a prominent Boston patent attorney.

The Bells had two daughters, Elsie and Marian, better known as Daisy. They also had two sons, both of whom unfortunately died at a young age. One of the boys was born in Rockport. Here is how that happened.

In the summer of 1881, Bell was called to Washington. He was asked to develop a metal detection device quickly. It was to be used to find the bullet lodged in President Garfield, whom an assassin had shot. Although his detector was not able to help the president, it was used successfully for many years, before X-ray machines, to find bullets lodged in victims and save their lives.

While Alec was working in Washington, Mabel, who was pregnant, was living with their daughters in Cambridge. In the middle of August, that summer of 1881, Bell came home to take his wife and the girls for a seaside vacation. They stayed at a house on Granite Street in Rockport across from Pigeon Cove. The house is still there and is marked with a sign proclaiming Bell's visit.

On August 15, a few days after they arrived, Mabel gave birth prematurely to a son. Unfortunately, the boy died from breathing problems three hours after his birth. The event was noted in a telegram Bell sent from Pigeon Cove to his father. This tragedy prompted the inventor to undertake a project to develop an artificial breathing apparatus for children, including his own son. *He eventually succeeded in coming up with a design for the first "iron lung".*

The Bells enjoyed spending summers by the sea. A few years later, during the summer of 1884, they vacationed in a house on the Magnolia coast. The exact location is not precise. It was described as "in sight of Norman's Woe". This would place it near the current location of Hammond Castle on Hesperus Avenue.

Alexander Graham Bell is one of the most famous and influential inventors of the 19th century, alongside Thomas Edison. The North Shore of Boston played an essential role in his life. Salem was where he started his technological developments that led to the invention of the telephone. And Rockport was where he began work on his creation of the first iron lung, as a way to mourn the loss of his son.

Update of the article

Beyond his groundbreaking work on the telephone, Alexander Graham Bell made significant contributions to many fields of science and technology. Much of this later work was conducted in his Canadian home and laboratory. They were located in Baddeck, Nova Scotia, on the shores of the Bras d'Or Lakes.

Known as Beinn Bhreagh (Gaelic for "Beautiful Mountain"), this sprawling estate served as both a summer residence and a research hub for Bell and his family.

Here, Bell conducted pioneering experiments in aviation, hydrofoils, and other technologies, collaborating with his wife, Mabel, and a team of dedicated assistants. The scenic and tranquil setting of Bras d'Or Lakes provided Bell with inspiration and isolation for innovation. At the same time, the estate itself became a bustling center of scientific exploration, producing remarkable achievements, including some of the earliest powered flight experiments in Canada.

Bell was a founding member of the *Aerial Experiment Association* (AEA) in 1907, a group dedicated to advancing powered flight. Under his leadership, the AEA developed several innovative aircraft, including the *Silver Dart, which became the first powered aircraft to fly in Canada in 1909.* Bell's interest in aerodynamics and his experiments with tetrahedral kites also contributed to the early understanding of lift and flight stability.

Bell was also a *pioneer in medical technology*. His son's death caused him to invent the iron lung. Motivated by his mother's and wife's deafness, he developed numerous devices for the hearing-impaired. His "*audiometer*," an instrument for measuring hearing acuity, was an early diagnostic tool for detecting hearing loss.

Bell also invented the "*photophone*" in 1880, a device that could transmit sound on a beam of light, *a concept that anticipated the principles of modern fiber-optic communication.* Although the photophone was not commercially successful in his time, it laid the groundwork for future innovations in optical communication.

In addition to his work in telecommunications and aviation, Bell was an advocate of scientific research in many other fields. He conducted *experiments in marine engineering, designing hydrofoil boats* that set speed records in the early 20th century.

He also *studied animal breeding*, developing a herd of superior sheep at his Nova Scotia estate where he applied principles of selective breeding. Bell's diverse interests and inventive spirit extended far beyond the telephone, making him one of the most versatile scientists and inventors of his era.

The Remarkable Hammonds of Gloucester

Submitted to the Gloucester Daily Times but not yet published.

Cape Ann has been home to many exceptional people. None have been more remarkable than two men, father and son, both named John Hays Hammond. While they were not born here, each was a successful entrepreneur who built an elegant estate along the western harbor front and spent their last years here in Gloucester.

Everyone knows Hammond Castle on Hesperus Avenue in Magnolia. It was the medieval-style dream house of John Jr., affectionately known as Jack by his family. Less well known was the showplace estate of John Sr., located on Western Avenue, just north of Freshwater Cove. Many will remember that this is the villa deceptively purchased by the Unification church in 1978, along with several other parcels along Gloucester's waterfront. Reverend Sun Myung Moon, like Jack Hammond, had a dream. He wanted to build a U.S.-based maritime enterprise to support his church. But that is another story.

John Sr. was born into a life of high adventure. His father, Richard, was a West Point graduate who became a Major in the Army and fought in the Mexican-American War, 1846-1848. In 1849, gold was found at Sutter's Mill in Coloma, California, about sixty miles northeast of San Francisco. This was the start of the great Gold Rush, and it excited Richard Hammond to think about it.

The Major, seeking his fortune, brought his wife, Sarah Hays, to San Francisco. He became a prospector and left her behind in the city-by-the-bay while he began his quest in the gold fields. It was several years later, in 1855, when John Sr. was born to the Major and Sarah. San Francisco had grown dramatically and was populated with hordes of rough men who came to work the mines. It was a lawless time, especially in the section of the city known as the Barbary

Coast. Criminals and gambling establishments populated the area. The young Hammond Sr. grew up in the Wild West of old San Francisco, which fueled his lifelong need for adventure and risk-taking.

John was a very bright young man. When it came time for him to attend college, he chose *Yale University* and its Sheffield Scientific School. Growing up during the gold rush, the young man had become very interested in the science and technology of mining. After receiving his Yale degree, he headed to Europe to study mining at the *Royal School of Mines in Freiberg, Germany*. He spent three years there and became an expert in the field. It was also there that he met his future wife, Natalie Harris, who was born in Vicksburg, Mississippi. She was also a student in Germany. She was studying music in Dresden, about an hour away by train.

After a few jobs as a mining engineer with American companies, John headed to South Africa with his family, including his young son, Jack, to evaluate the gold mines in Transvaal for American investors. His big break came when Cecil Rhodes, the British owner of many gold and diamond mines in Africa, hired him. Hammond rose quickly in the organization to the point where he was, at the age of forty, in charge of all of Rhodes' mines. And he made each one a significant financial success.

But his good fortune was not to last. John was caught up in the local politics that led to a war between the British and the Dutch settlers, known as the Boers. The Boers captured Hammond, and for his role in the conflict, he was sentenced to death. Fortunately for him, his boss, Cecil Rhodes, was able to get his release by paying fines to the Boer government.

These experiences, especially his success in the mining business, gave John worldwide fame. When he returned to the U.S., he was only in his mid-forties and became a highly paid consulting engineer for mines. He made millions of dollars working for wealthy owners to evaluate and improve their mining operations.

But his fame opened other doors for him. He had gone to Yale with William Howard Taft, who was the President of the United States (1909-1913). This friendship led to Hammond moving to Washington to become a trusted advisor and emissary for Taft. John travelled extensively around the world representing Taft and America's interests.

Another profitable relationship for John was with Frederick Russell Burnham, owner of an oil exploration company. The company struck a primary oil site near Carson, California, making Hammond even wealthier.

John Sr. had a profound impact on many lives around the world, both in business and politics. As a result, in 1926, when Hammond was 71 years old, a group of his friends and associates formed *"The Company and Friends of John Hays Hammond"*. They sponsored eleven dinners in his honor in various major cities around the world, including Manhattan, London, Paris, Tokyo, and Manila.

Amazingly, more than 10,000 of the rich and famous of the time wrote letters of tribute to the man, including Presidents and wealthy industrialists. The event was so noteworthy that John Hays Hammond appeared on the May 10, 1926, cover of Time Magazine, with a biographical article about him titled "*Unique*".

John Hays Hammond, Jr., was as remarkable as his father. Like his Dad, Jack was born in San Francisco. The year was 1888, and John Sr. was working as a mining engineer in the area. But at the age of five, the family moved to the Transvaal in South Africa. John Sr. had an assignment there to evaluate mines for American investors. In 1898, the family moved to England after John Sr. was rescued from the Boers by Cecil Rhodes. *They remained there for a couple of years before returning to the U.S.*

These were important years for Jack. He was ten years old and became fascinated with medieval history. This fascination later led him to build his castle on the harbor front of Gloucester.

The family returned to the U.S. in 1900. John Sr. wanted to inspire his son, so he took him on a business trip, at the age of twelve, to meet Thomas Edison. The site was Edison's laboratory in West Orange, New Jersey. This changed the boy's life. With the boy's many intelligent questions, *Edison took an interest in Jack.* He became a mentor to the young man, and they remained in touch thereafter.

Like John, Sr., Jack chose to attend Yale University and Sheffield School. His goal was to become an inventor like Edison, and the science and technology education he received at Yale would serve him well. While there, Jack became especially interested in the study of a new technology, radio.

Once again, his father's broad contacts put him in touch with another technology giant, *Alexander Graham Bell.* And again the young man was able to impress the great inventor of the telephone with his intellect and interest. Bell also became a lifetime counselor to Jack. *It seems incredible that Jack would have as his mentors two of the most important inventors in the history of the world.*

And so he embarked on his lifetime of creating new technology. He built his research operation on his father's Gloucester estate. *It was called the Hammond Radio Research Laboratory*. Later, in about 1929, he moved it to his newly constructed castle in Magnolia.

It was in Gloucester that *he invented over 400 devices, mostly in the fields* of radio control and naval weaponry. He became known as the Father of Radio Control, and *his inventions are key to today's ubiquitous drone aircraft and missile guidance systems.*

John Sr. died in Gloucester in 1936 at the age of 81. Jack was living in his Magnolia castle, but he died in New York City at the age of 76 while he was attending a Board meeting of R.C.A.

Cape Ann's history is enriched by the remarkable lives of John Hays Hammond Sr. and Jr., two men whose adventurous spirits, pioneering achievements, and grand estates left an indelible mark on Gloucester's landscape and legacy. From the gold fields of California to the scientific halls of Yale, from South African mines to the towering walls of Hammond Castle, their stories are woven with ambition, brilliance, and global influence. Yet, at the heart of their journeys was the coastal community they had chosen as home. Their contributions, both in industry and innovation, continue to inspire, reminding us that even in a quiet harbor town, greatness can take root and thrive.

Update of the article

The impact of John Hays Hammond Jr.'s developments on today's technology - John Hays Hammond Jr., often hailed as the "Father of Radio Control," significantly influenced the development of modern unmanned systems, including drones and missile guidance technologies. His pioneering work in the early 20th century laid the foundation for the sophisticated remote-controlled systems utilized in today's military and civilian applications.

One of his notable achievements was the installation of radio control gear on the obsolete U.S. Navy battleship USS Iowa in 1922, transforming it into a remotely controlled target ship. This experiment demonstrated the feasibility of *using radio waves to control large vessels without onboard crews*, a concept that would evolve into modern unmanned systems.

Hammond's innovations extended to anti-jamming systems and multi-channel radio systems, enhancing the reliability and security of remote communications. These advancements were crucial during World War II, as various nations developed radio-controlled weapons and vehicles. For instance, the German military employed radio-controlled missiles and glide bombs, while the Soviet Red Army utilized remotely controlled tanks. The United Kingdom and the United States also experimented with radio-controlled aircraft for training and combat purposes.

The principles established by Hammond's early work are evident in today's drone and missile technologies. Modern unmanned aerial vehicles (UAVs) and unmanned combat aerial vehicles (UCAVs) rely on secure, multi-channel radio communications for remote operations. Similarly, contemporary missile guidance systems utilize advanced forms of radio control to target their objectives with precision. Hammond's emphasis on secure and reliable remote-control mechanisms has been instrumental in the evolution of these technologies.

The Russia-Ukraine war has seen an unprecedented surge in the use of drones, fundamentally transforming modern warfare. Both nations have rapidly expanded their drone fleets, deploying a wide array of unmanned systems for reconnaissance, precision strikes, and electronic warfare. Ukraine, facing manpower shortages, has turned to mass-producing drones, aiming to manufacture up to four million annually, ranging from small FPV (First Person View) models to larger aerial and naval drones.

Russia, in response, has increased its drone production, including the development of AI-powered systems and the establishment of facilities to manufacture drones such as the Shahed-136. These drones have been employed extensively, with Russia launching over 100 drone strikes in a single night, targeting Ukrainian infrastructure despite international ceasefire efforts. The widespread use of drones has not only altered battlefield tactics but also underscored the strategic importance of unmanned systems in contemporary conflicts.

In summary, John Hays Hammond Jr.'s groundbreaking work in radio control has had a lasting impact on the development of unmanned systems. His innovations laid the technological foundation for modern drones and missile guidance systems, underscoring his pivotal role in the history of remote-controlled technology.

Oppenheimer and the atomic bomb

First published in the Gloucester Daily Times on September 19, 2023

Christopher Nolan, one of Hollywood's premier Directors, has written and presented a highly praised film about the life and work of J. Robert Oppenheimer, the man who led the Manhattan Project during World War II. The 2023 film, "*Oppenheimer*," has, once again, raised many questions about the ethical propriety of designing and building the first atomic bombs. Should we have created such civilization-destroying weapons? Should we have used them against Japan to bring the war to an end? Should we continue to have such weapons in our arsenal?

The answer to the first question was obvious to the men and women behind the Manhattan Project, including Oppenheimer. They had no choice. If the United States had not succeeded first, then the Germans or the Russians would have. *The results would have been the deaths of millions of American civilians, the loss of the war, and the end of civilization as we know it.*

Oppenheimer had a complex and nuanced attitude toward the building and use of nuclear weapons. He believed that the atomic bomb was a necessary evil, but he was also profoundly aware of its destructive potential and the ethical implications of its use. So, on the one hand, he was very concerned about the role he played in creating such horrific weapons. Yet, for the rest of his life, he defended their use to end World War II and prevent World War III from starting.

Oppenheimer was not alone in his mixed feelings. President Harry Truman wrestled for a long time about giving the order to use the weapons on Hiroshima and Nagasaki. The Japanese military appeared unwilling to surrender unconditionally to end the war and was determined to fight to the last person for their homeland. So, for Truman, it was either to invade the Japanese mainland or to use horrific weapons. The U.S. military experts estimated that, *in the invasion, we would lose up to one million American soldiers, and about five million Japanese soldiers and civilians would also die.*

Tragically, about 200,000 Japanese did die from the atomic bombs, but that total was less than 5% of the 5 million Japanese lives that otherwise would have been lost with an allied invasion of their homeland.

After the war, many religious leaders from around the world made public statements against the use of nuclear weapons. The Catholic Church, for example, condemned the use of nuclear weapons as immoral and unjust. In 1963, Pope John XXIII issued an encyclical called "Pacem in Terris" (Peace on Earth), in which he wrote that "*the use of nuclear weapons must be condemned as a crime against humanity and against God*". He also called for a ban on nuclear testing and the destruction of all nuclear weapons. The Dalai Lama made similar statements.

For me, the moral questions about the use of nuclear weapons are very personal. I did not contribute to the Manhattan Project, but I worked with a group of men who did. In August 1942, at the start of the Project, these men were Professors and students at the Massachusetts Institute of Technology. They had backgrounds in electronics and highly sophisticated photographic technologies and were brought to Los Alamos, New Mexico, to apply their expertise to the Project.

After the war, the Atomic Energy Commission asked the M.I.T. group to continue its involvement. As a result, Harold "Doc" Edgerton, Kenneth Germeshausen, and Herbert Grier formed a company, Edgerton, Germeshausen & Grier (E.G.&G.), based in the Boston area with operations at the Nevada Test Site and the infamous Area 51.

When I joined their company as a physicist in 1965, the Cold War was at its height. I worked on simulating the effects of nuclear weapons on materials. The information from these experiments was used to design missile warheads that were more impervious to an enemy's use of another nuclear weapon to destroy our incoming warhead. This technological adaptation made the enemy less likely to use their weapons against us because they knew they could not stop the assured destruction of their country if they did.

Like Oppenheimer, I am unhappy for having played a role in the development of these dreadful weapons. However, like him, I feel positive about the fact that their existence is under our control, preventing others from ever using them in anger against us or anyone else. This has been mainly due to the policy of Mutually Assured Destruction (M.A.D.). It has been successful since the United States built and kept up an overwhelming arsenal of these weapons and their delivery systems, so no other country has dared to attack us.

This strategy relies on all other countries that possess these weapons to be rational actors. The leaders of Russia and China, thus far, have been. We cannot, however, be as sure about North Korea or Iran. These countries do not yet possess an adequate nuclear force to pose a threat, but they could do so in the near future. So, we need to have additional deterrence just in case.

Beginning with President Reagan's Strategic Defense Initiative, also known as the "Star Wars" program, the U.S. has been developing technologies that can protect against a nuclear attack. Since 2004, we have deployed both land-based and sea-based anti-ballistic missile systems. They have been evaluated many times and have been successful in knocking down incoming warheads. They are not, however, perfect and have failed to hit a few of the incoming missile test targets. So, more needs to be done.

The ultimate systems, part of President Trump's "Golden Dome", are under development. They are high-powered, land and sea-based lasers that can track and destroy incoming warheads at the speed of light. Hopefully, when these systems are perfected and deployed, we will be able to rest more easily.

Update of the article

The latest developments in the use of high-power lasers to shoot down incoming ballistic missiles - The United States is advancing the development of high-power laser systems to enhance missile defense capabilities, with a focus on integrating directed-energy weapons (DEWs) across various military platforms. This approach would be an element of "The Golden Dome" program proposed by President Trump.

The U.S. Navy has made significant strides on the *High-Energy Laser with the Integrated Optical Dazzler and Surveillance* (HELIOS) system. Developed by Lockheed Martin, HELIOS is a 60-kilowatt laser weapon designed to intercept drones, fast-attack craft, and missiles. It has been installed on the USS Preble (DDG-88) and is integrated with the Aegis Combat System, enhancing the ship's defensive capabilities.

In addition, the Navy is preparing for intercept tests with the 300-kilowatt *High Energy Laser Counter Anti-Ship Cruise Missile Project* (HELCAP), aiming to provide a cost-effective solution against the increasing threat of drone and missile attacks.

The U.S. Army is also progressing in deploying laser weapons. It has officially deployed high-energy lasers overseas to counter enemy drones, marking a significant milestone in the operational use of directed energy weapons (DEWs). Furthermore, the Army is evaluating additional 50-kilowatt platforms for short-range air defense, seeking to enhance its capabilities against aerial threats.

The Missile Defense Agency (MDA) is revisiting airborne laser systems, focusing initially on using low-powered lasers for tracking, with plans to develop higher-powered systems for missile interception. This cautious approach follows previous challenges and technical problems in airborne laser development, with the aim of building effective airborne missile defense capabilities incrementally.

Despite these advancements, challenges remain in the widespread deployment of laser weapons. The U.S. Navy, for instance, faces criticism over the slow integration of directed energy weapons on warships, with officials expressing frustration over the lack of progress despite significant investment. Nevertheless, the continued development and testing of high-power laser systems underscore the U.S. military's commitment to enhancing its missile defense architecture through directed-energy technologies.

Should we apologize for Hiroshima?

First published by the Gloucester Daily Times on May 26, 2016

Following the Group of Seven economic summit meeting, to be held this week in the Ise-Shima region of Japan, President Obama will visit the city of Hiroshima and be accompanied by Japan's Prime Minister, Shinzo Abe. He will be the first sitting U.S. President to make such a stop. He said recently in an interview with the Japanese press that he will not apologize for the U.S. dropping the atomic bomb on that city. However, it's not that he didn't want to apologize. The Japanese government insisted that he not do so.

We know this because in September 2009, during his series of apology tours, Obama was preparing for a visit to Japan and secretly raised the idea of making an apology to the Japanese for Hiroshima and Nagasaki. The Japanese response was contained in a secret cable sent to Secretary of State Hillary Clinton, dated September 9, 2009, that was released by WikiLeaks in 2011.

The cable reported that Japan's Vice Foreign Minister, Mitoji Yabunaka, told John Roos, the U.S. Ambassador to Japan, "the idea of President Obama visiting Hiroshima to apologize for the atomic bombing during World War II is a 'nonstarter.'" The Japanese were concerned about the internal political ramifications of such an apology. So, Obama did not visit Hiroshima on that November 2009 trip, but he did make world headlines when he became the first U.S. President to make a full bow to the Emperor of Japan.

But does the United States have anything to apologize for? *Not when you take the development and use of the first atomic bombs into the historical context of the events that led the United States to undertake the Manhattan Project, and then the decision that President Truman made to use the weapons.*

In the fall of 1938, it was clear that the Third Reich was on a course of expansionism that could ultimately lead to a World War. Earlier that year, the Nazis had occupied the Sudetenland, a region of Czechoslovakia, with the pretext of aiding German nationals living there. Then in March of 1938, under the threat of invasion, the Anschluss (Annexation) of Austria to Nazi Germany was enacted.

During that same period, scientists in several countries were working on physics experiments, using neutrons to bombard uranium. They were trying to understand the interaction of the neutrons with the uranium atoms. Two German chemists, Dr. Otto Hahn and Dr. Fritz Strassmann, used a new technique to analyze the results and found an astonishing and unexpected result. They found that the interaction resulted in the presence of barium atoms along with the release of high-energy neutrons.

Now, barium is another element whose atomic number is 56, compared to 92 for uranium. Hahn and Strassmann were not theoretical physicists, so they needed help in explaining what was happening. They turned to Dr. Lise Meitner and her nephew, Dr. Otto Hirsch, two former German scientists who had escaped to Denmark under threats from the Nazis. The two nuclear physicists were intrigued and set to work to understand the results. They developed a nuclear model that aligns with the experiment. They concluded that the uranium atoms were being split into two pieces, creating barium, *along with the release of a huge amount of energy in the form of highly energetic neutrons.*

When they published the results in a physics journal, the scientific world suddenly realized that this phenomenon had the potential to make a weapon with unimaginable, destructive power.

In the United States, physicists saw that potential and worried that the Nazis would develop a bomb first. These scientists included Albert Einstein, who was living in the United States and was a professor at Princeton University. It was Einstein's Theory of Relativity that helped to explain the great amount of energy that was released from the fission of the uranium atom. He signed a secret letter to President Roosevelt urging him to undertake the development of the bomb. The result was the super-secret Manhattan Project.

The Nazis did indeed also launch their secret program, Uranprojekt, to develop a nuclear weapon. Fortunately, their effort did not go as well. Many of the most competent scientists, engineers and mathematicians had left Germany starting in 1933, concerned with the Nazi interest in purging Jewish scientists from Academe. In addition, the Allies had been successful in launching a black ops action that further slowed their efforts. By the end of the war in 1945, the Nazi nuclear weapons program was still ongoing, but they were far from achieving success.

What would have happened if the Nazis were successful before the United States? The outcome of the war would have, most likely, been very different. It is easy to imagine scenarios where the Nazis would demonstrate the power of for such a weapon on a Western city, say in England. This could have convinced the Allies that they had no choice but to surrender to the Third Reich or risk massive destruction. So, the United States' preemptive development of a nuclear weapon literally saved the civilized world.

Meanwhile, the Allies were also waging a war with the Japanese. In 1943, the enemy was being forced back across the Pacific. The Marines fought bloody battles from atoll to island across the whole region. It was a difficult and deadly war with highly trained Japanese soldiers dug in across the islands and willing to fight to the death.

By the summer of 1945, the Japanese were finally cornered in and around their homeland. But they showed no sign of capitulating. So, the Allies planned to invade their islands with large-scale forces that they called Operation Downfall. 700,000 Allied troops were being readied to land on Kyushu, the southernmost of the big four Japanese islands. *This would have been an operation even larger than D-Day*. And the U.S. military expected a bloody slaughter with high casualties on both sides.

It was in this context that President Harry Truman had to decide. Either drop atomic bombs on Hiroshima and Nagasaki or authorize Operation Downfall.

Today, many people say that it was a mistake to drop the bombs on those two cities, killing about 200,000 Japanese civilians. But, to Truman and his advisors at the time, it was a "no-brainer". 36 million people, military and civilian, had already died in the war in the Pacific. The U.S. had lost 111,000 troops and 253,000 wounded. The Japanese had demonstrated their willingness to die rather than surrender, and Truman didn't want to take the risk of another 500,000 casualties (both Allies and Japanese) as was estimated at the time. So, he issued the order for the Enola Gay to embark on her deadly missions.

Polls of the American people taken at the time showed overwhelming support for Truman's decision. And all of those hundreds of thousands of families who might have lost their sons in the invasion thanked President Truman for his courage. As the years passed and memories of the horrific battles fought in the Pacific faded, more people expressed disapproval in later polls; however, the majority still approved of the decision as the necessary action for the time and circumstances.

Update of the article

The global perspective on the United States' use of atomic bombs during World War II has evolved significantly over the decades, reflecting a complex interplay of historical memory, ethical considerations, and geopolitical contexts. Initially, the bombings of Hiroshima and Nagasaki were largely viewed within the U.S. as necessary actions to hasten the end of the war and minimize further casualties. However, as time has passed, *international attitudes have shifted,* with increasing scrutiny and debate over the moral and legal justifications for deploying nuclear weapons against civilian populations.

In Japan, the bombings are remembered as profound national tragedies. A 2015 survey indicated that 79% of Japanese respondents believed the bombings could not be justified, highlighting a deep-seated opposition to nuclear warfare.

This sentiment is further underscored by the activism of hibakusha (atomic bomb survivors), who have played a pivotal role in advocating for nuclear disarmament. Organizations like Nihon Hidankyo, which received the Nobel Peace Prize in 2024, have been instrumental in bringing global attention to the humanitarian consequences of nuclear weapons.

European opinions also reflect a critical stance. A 2025 international survey revealed that majorities in countries like Germany (81%), Italy (78%), and Spain (75%) considered the bombings morally unjustified. *In contrast, the United States displayed a more divided perspective, with 38% viewing the actions as morally justified, 35% opposing them, and 26% uncertain.* This divergence underscores differing historical narratives and the complexities of wartime decision-making.

In other parts of Asia, perspectives are shaped by unique historical experiences. In South Korea, while some view the bombings as contributing to liberation from Japanese colonial rule, there is also criticism due to the significant number of Korean victims in Hiroshima and Nagasaki.

Overall, the global attitude toward the U.S. use of atomic weapons during World War II is characterized by a growing emphasis on the humanitarian impact and ethical implications of nuclear warfare. While some still argue that the bombings were a necessary means to end the war swiftly, there is an increasing consensus on the need to remember the profound suffering caused and to work toward a future free of nuclear weapons.

Meeting with the Jumping General

First published by the Gloucester Daily Times on January 13, 2016

I remember the morning well, even though it happened in the early eighties. I was waiting for the General to meet me in my office. I had met him a few times before, but this time was to be special. He was going to call his friend, General Matthew Ridgway. I was to be introduced to the famous World War II commander of the XVIII Airborne Corps. General Ridgway was a member of the Board of Directors of a particular company, and I wanted to meet with the Chairman of that firm to discuss a business matter. I had hoped that Ridgway could arrange the visit for me.

Eventually, the tall, slender, ramrod-straight General James Gavin entered my office. He had been the commander of the 82nd Airborne Division during World War II and had jumped with his men into several battles, including Sicily, Normandy, and the Battle of the Bulge. Since he had more combat jumps than any other general officer, he was termed by the press "The Jumping General".

We started to talk. I remember him asking me if I had thought of joining the Airborne and becoming a trooper. He wasn't serious since I was already a 30-something management consultant at Arthur D. Little, Inc. (ADL) while the General was the Chairman of the Board. He had been the Ambassador to France under President Kennedy, and the Chairman of ADL for many years, yet his mind was still rooted in his time as the 82nd's commander during wartime. And well it should have been.

Gloucester has a strong Marines JROTC program, and many of our teens look to the military as a place of honor to serve their country. General Gavin's life story could be an inspirational one for them. His character was molded in the military, and his strong, decisive approach to problems guided him throughout his life.

His beginnings were rather humble. At age two, Gavin was placed in a Brooklyn orphanage by his alcoholic mother, Katherine Ryan. Two years later, Martin and Mary Gavin adopted him. Martin was a hardworking coal miner from Mt. Carmel, Pennsylvania. The Gavins were poor, so Jim left school after the eighth grade to work and help support his family. At 17, he saw no future for himself in Mt. Carmel, so he ran away to New York to join the Army. Even though he was underage, the recruiter found a way to get the young man into the service. For his first assignment, he was stationed with an artillery group in Panama.

The young Private was extraordinarily bright and an intense reader. During his off hours from his military duties, he read extensively in the camp library about the Civil War and World War I, focusing on famous Generals. He became excited about the idea of commanding men in battle and *decided that he wanted to go to West Point.* Encouraged by his superiors, who saw his potential, Corporal Gavin passed the exams and entered the Academy at age 19. He had to work harder than the other cadets because he didn't have a formal high school education. Yet, he did well and graduated from the Academy in 1929.

General Gavin's exploits during World War II were nothing short of amazing. He described them in his book, "On to Berlin". I've read that book a few times and always find myself in awe at the heroic actions of Gavin and his troopers.

After graduation from West Point, the young Lieutenant rapidly rose in the Army. He had a few jobs in Infantry units and then went on to teach at the Military Academy. It was the late 1930s, and the ever-curious young officer became interested in the possibilities for using air-deployed forces. The Germans were among the first to develop airborne troops as a military weapon. He studied their tactics on one of their airborne assaults during the invasion of Belgium in May 1940. Gavin was intrigued and volunteered in April of 1941 for a new airborne unit that was being formed. *That's when he first met General Ridgway.*

Gavin, the bright, energetic, personable young Captain, impressed his commanding officers, and he was given the responsibility for developing the tactics and basic rules of airborne combat. He wrote the manual that the Army would then use in its training. *As a result, he became recognized as the architect of the emerging airborne tactics.* This fueled his rapid rise in an army that was now clearly headed for war. He soon rose to the rank of Colonel and had command of the 505th Parachute Regiment of the newly formed 82nd Airborne Division, under the command of General Ridgway.

After months of exhausting training, Gavin and his men were ready for their first mission. Ridgway selected Gavin's regiment for a critical assignment in the invasion of Sicily. Gavin accompanied his men and made his first combat jump onto the island. The mission was to support General Patton's landing of his troops onto the southern beaches. The 505th jump went down in history as the first regimental-sized airborne operation by any Army during war.

His success further enhanced Gavin's reputation. Despite significant obstacles from the weather and the enemy, he led his troops in a series of engagements that contributed to the campaign's success. Many of his men were lost in the heavy fighting, and Gavin mourned every one of them. Molded by their commander, they truly were a "band of brothers."

As a result of his actions, Gavin was promoted to Brigadier General under the recommendation of General Ridgway. He was made Ridgway's Assistant Commander of the 82nd. *Now at 37 years of age, Gavin was the youngest General in the Army since George Armstrong Custer.*

Gavin soon became the commander of the 82nd when Ridgway was promoted to the command of the XVIII Airborne Corps. The young General went on to plan extensively for airborne missions in Normandy and the Netherlands. He participated with his men in the execution of the combat jumps. One of these was a joint assault with British forces, the famous Operation Market-Garden. It was also known as "a bridge too far" because it did not go as planned. It was portrayed in a book and a movie of the same name.

For his courage and gallantry in extreme combat conditions during the war, General Gavin won many awards. His first was a Distinguished Service Cross, second only to the Medal of Honor. General Ridgway recommended it for Gavin's actions in the Sicily invasion. He went on to receive another Distinguished Service Cross, as well as two Silver Stars and the Purple Heart. The British and French governments also decorated him with some of their highest honors. When the War ended, he had two stars as a Major General and was still only thirty-eight years old.

General Gavin came to my office that day, so many years ago, to help me with a business contact. The contact was made, and I had the meeting I was seeking. However, more importantly to me, I had the opportunity to spend some time with one of our greatest heroes of World War II.

Update of the article

After World War II, Generals James M. Gavin and Matthew B. Ridgway continued to serve in significant military and public roles beyond their positions on corporate boards.

James M. Gavin, after World War II, remained active in the U.S. Army, eventually becoming the *Army's Chief of Research and Development.* In this role, he was instrumental in modernizing the Army, advocating for the integration of airborne mechanized infantry and the development of the "airmobile" concept, which *emphasized the use of helicopters for rapid troop deployment.* This strategy was extensively used during the Vietnam War.

In 1961, President John F. Kennedy appointed Gavin as the U.S. Ambassador to France, leveraging his wartime experiences and relationships to improve diplomatic relations during a period of tension. Gavin served in this role until 1962.

Later, he became a vocal critic of the Vietnam War, aligning with other former military leaders who questioned the U.S. involvement. His anti-war stance led to political engagement, including being approached to run against President Lyndon B. Johnson in the 1968 Democratic primaries, though he declined.

Matthew B. Ridgway, after World War II, held several key military positions. During the Korean War, Ridgway was appointed commander of the Eighth Army, where he revitalized U.S. forces and led successful operations. When President Truman relieved General Douglas MacArthur, General Ridgway succeeded him as commander of the United Nations Command in Korea. In 1952, he followed General Dwight D. Eisenhower as Supreme Allied Commander Europe, and in 1953, he became the U.S. Army Chief of Staff. *Ridgway was known for his opposition to overreliance on nuclear weapons, advocating instead for strong conventional forces.*

After retiring from active duty in 1955, Ridgway continued to be engaged in public service and policy. He served as chairman of the board of trustees of the Mellon Institute and participated in various Pentagon strategic study committees. Ridgway was also a founding board member in 1976 of the *Committee on the Present Danger,* promoting greater military preparedness during the Cold War.

Both Generals James M. Gavin and Matthew B. Ridgway exemplified a commitment to leadership and public service that extended far beyond their military careers. As innovators, diplomats, and outspoken advocates for principled defense policies, they shaped not only the U.S. Army but also the nation's strategic direction during a critical period of the Cold War. Their post-war contributions demonstrate that leadership is not confined to the battlefield but continues in the boardroom, the diplomatic arena, and the public sphere, leaving a legacy of courage, innovation, and integrity.

Transformation into an American patriot

First published in the Gloucester Daily Times on August 28, 2024

One of my neighbors, Yuri Tuvim, recently passed away. I met Yuri several times over the years when we bumped into each other during walks around Annisquam Village. But I didn't know anything about him, except that he was a conservative writer and activist. Then, seven years ago, we had the opportunity to discuss the matter in more detail. I learned about his inspiring story. His saga begins with him living in Russia as a member of a communist youth organization, where he firmly believed in the Communist dream. But later, he is transformed into a leading Soviet dissident in the movement that ultimately contributed to the decline and fall of the Soviet Union.

Yuri Richardovich Tuvim was born in 1930 in the city of Ryazan, on the banks of the Oka River, one hundred and twenty-two miles southeast of Moscow. His father, Richard, was the chief engineer at a liquor manufacturing plant located in Ryazan. His mother, Sarah, was a dentist. As a result, the family lived relatively well in those days. The plant where his father worked provided a small but comfortable apartment for the Tuvim family.

The schools in Ryazan were good. Students at that time were still taught by teachers who were trained before the Russian Revolution. But underlying the whole education system was the Soviet program for indoctrinating the youth. Children were taught that capitalism was bad for the masses because the owners exploited the workers. They learned that the Soviet system is the future of mankind, where everyone is equal and receives what they need from the State. Yuri's parents didn't believe this, but they never spoke to him about the lies of Communism. *They understood that if Yuri said the wrong word in the classroom, the whole family would be in trouble.*

World War II brought hard times to Ryazan. The German Army advanced to within 16 miles of the city before being forced to retreat. Throughout the war years, food was very scarce, and everyone suffered. In 1944, it was normal for a boy of fourteen to become a member of Komsomol, the Communist youth organization, and so Yuri joined.

At one point during the war, Yuri remembers being cold, hungry, and dressed in rags while looking at a world map in his classroom. He thought, "There must be a boy like me in America who has it even worse than me because he is living under capitalism. How lucky I am to be living here in the Soviet Union." Later, Yuri recognized that he had been completely "brainwashed" by the Soviet indoctrination program.

The central turning point in Yuri's life came in 1948, at the age of 18. He was then living in Smolensk, about 220 miles southwest of Moscow. His father had been sent there to restore a liquor manufacturing plant that was destroyed during the war. Yuri spent his last year of high school in Smolensk and did very well. He graduated with a Gold Medal award for scholarship. This meant he could attend any university he wished without examination. He chose the Institute of Aviation in Moscow because he was interested in designing aircraft engines.

He soon found that being admitted to that Institute would not be possible. All Jewish students were automatically rejected. This was devastating to Yuri. For his whole life, he had been told that everyone was equal under the Soviet system, but now he learned that, because he was a Jew, he was turned down. *His deep-seated belief in the Soviet system was shattered.*

Yuri did get into a college of lesser interest to him, the Institute of Chemical Machine Building. Again, he graduated near the top of his class in 1953, but found that, because of the now rampant anti-Semitism in the Soviet system, he couldn't get a good job. This reinforced his disenchantment with the Soviets. These feelings intensified with his recognition of a wide range of other human rights abuses by the Soviets.

Many Russians shared this disillusionment, especially Russian Jews. After the death of Josef Stalin in 1953, members of the Moscow Intelligentsia started to find ways to protest the wrongs they saw in the Soviet system. They became known as "dissidents". It was a dangerous activity. Many were sent to labor camps, psychiatric hospitals, or even murdered.

In the summer of 1965, Yuri accompanied his brother-in-law, Dr. Akiva Yaglom, a famous mathematician, to a conference in Tbilisi, Georgia. Akiva introduced Yuri, now a thirty-five-year-old engineer, to Akiva's classmate and friend from Moscow University, *Andrei Sakharov*. Yuri didn't know it at the time, but Sakharov, the father of Russia's hydrogen bomb, *was about to become one of the most famous Soviet dissidents of the era.*

Because of the stimulating conversations between Tuvim and Sakharov, the older scientist drew Yuri into his circle of dissidents. They became close friends and colleagues right up to the time that Yuri was forced to leave the Soviet Union.

Yuri became involved in many dissident activities. Hedrick Smith, a well-known New York Times correspondent, lived in Moscow during this period. Yuri collaborated anonymously with Smith on the journalist's book," The Russians", which turned into a major bestseller.

Finally, in 1975, Yuri had to leave his family and emigrate from the Soviet Union. Not because he wanted to, but because he was under threat of imminent arrest by the KGB. The KGB identified him as a danger to the state for helping to smuggle written materials out of the country.

Yuri, with great difficulty and trepidation, obtained an exit visa from Israel. Most people who had Israeli visas were not allowed to emigrate. They became known as Refuseniks. Fortunately, the KGB finally allowed Yuri to leave because they wanted to reduce the circle of friends around Sakharov.

Yuri eventually came to the U.S. and found a job as an engineer in Massachusetts. He married Pippy Giuliano of Annisquam and thrived in his new country. Knowing the hollow promises of the socialist system, he became a conservative activist here. He was committed to the idea of individual freedom and minimal government intrusions into citizens' daily lives, and he wanted to encourage others to have the same beliefs.

He contributed articles to newspapers and hosted a talk radio show during the 1980s. He was also a regular commentator on WEEI during the same period. The Gloucester Times published numerous letters written by Yuri.

Yuri Tuvim faced more hardship and danger than most of us will ever know. His story is a testament to the human spirit's resilience and the power of individual conscience. His unwavering commitment to freedom and justice serves as an inspiration to those who fight for a better world.

Update of the article

My experiences in the fading Soviet Union - In 1989, I traveled to the Soviet Union, specifically Moscow and Leningrad (now St. Petersburg), on a business assignment. My firm, The Winbridge Group, had been hired as a consultant by Spalding, the renowned American sports equipment manufacturer. My mission was clear: to evaluate whether the declining Soviet Union could serve as a low-cost manufacturing hub for products such as soccer balls and basketballs.

However, my findings quickly dispelled any optimism. During my assessment, it became evident that Soviet manufacturing lacked the quality control and precision necessary for producing high-standard sports equipment. The factories I inspected were burdened with outdated machinery and inefficient production processes, and the workforce, though skilled in some respects, was hampered by a lack of modern training and motivation. As a result, I advised Spalding against pursuing any manufacturing partnerships in the Soviet Union.

During my visit, the host was the Head of the Soviet government's sports equipment manufacturing organization, Glavsportprom. It served as the central state agency responsible for designing, manufacturing, and distributing sports equipment and apparel throughout the USSR. Operating under the broader Soviet sports bureaucracy, Glavsportprom played a crucial role in supporting both elite athletes and mass physical culture initiatives.

Established to centralize and streamline sports equipment production, Glavsportprom ensured that Soviet athletes had access to standardized, state-approved gear. This included equipment for various sports, such as boxing, bicycling, hockey, skiing, and a wide range of other athletic activities. These products were not only used domestically but also symbolized the USSR's commitment to excellence in sports on the international stage.

Glavsportprom's operations were part of a larger Soviet system designed to promote physical culture and sports as tools for ideological education and international prestige. By providing the necessary infrastructure and equipment, Glavsportprom contributed to the USSR's success in various international competitions, including the Olympics.

While in the country, I had many opportunities to observe what life was like for the average person in the Soviet Union. After more than seventy years under the Soviet system, it was clear how far behind the average Western country the Soviets were.

I stayed in their hotels, ate in their restaurants, shopped in their stores, visited private families in their apartments, and interacted with people where they worked. My overall impression was that the buildings, facilities, infrastructure, and people were drab, rundown, and demoralized.

A sense of struggle and uncertainty marked life for the average person in Moscow and Leningrad in 1989. The Soviet Union was in decline, and its citizens felt the impact of a stagnating economy. Long lines for basic necessities like bread, milk, and meat were a common sight, and even though stores were state-run, their shelves were often sparsely stocked. Citizens relied heavily on government rationing, and access to Western goods was extremely limited, available only to those with hard currency or connections to the black market.

Housing conditions were another significant challenge. Most people lived in cramped communal apartments (kommunalkas), where multiple families shared kitchens and bathrooms. Public transportation, in general, was extensive, overcrowded, and aging.

I must say, however, that I was very impressed with the Moscow Metro's underground system. It was clean and attractive, resembling a building from the turn of the twentieth century in terms of its style.

Despite these hardships, the people showed a remarkable sense of resilience. Cultural life continued to thrive in theaters, museums, and concert halls, providing a brief escape from the daily struggles. Many citizens still took pride in Soviet achievements, but beneath this pride lay an undercurrent of disillusionment as the realities of economic hardship and political instability became increasingly impossible to ignore.

As one of the few visitors from America at that time, many of the people I encountered were glad to meet me. I found most of them charming.

As part of a prearranged agreement, the head of Glavsportprom invited me to visit Artek, a renowned youth camp situated on the Black Sea in Crimea, a region of Ukraine. This visit was part of an initiative to explore the possibility of establishing a student exchange program between Artek and a group of young students from Waterbury, Connecticut, in the United States.

Artek was a storied institution, once celebrated as a premier summer camp for the Soviet youth group known as the Young Pioneers, but by the time of my visit, it had lost much of its former glory. The facilities, while functional, were visibly worn and somewhat drab.

During my inspection, I paid particular attention to the camp's medical facilities, a crucial factor in determining the safety and well-being of any American children who might participate in the exchange program. What I found was alarming. The medical facilities were rudimentary, reminiscent of the medical standards of the 1920s in the

United States. In an emergency, these facilities would struggle to provide adequate care for the visiting youngsters.

Interestingly, I learned that the same organization responsible for managing Artek, and which was my host for the visit, also administered a nearby enclave of dachas, private summer homes on the Black Sea, reserved for the highest-ranking members of the Soviet Central Committee. These dachas were luxurious compared to the stark simplicity of the Artek camp. As a guest, I was given a rare opportunity to tour these exclusive properties. I even had the chance to use the same sauna and dine in the same hall frequented by Soviet leaders, such as Mikhail Gorbachev and Nikita Khrushchev. Part of the property included a hunting reserve in the hills where they took me one night for an unforgettable barbecue.

Despite my concerns, the children's exchange went ahead, and students from the US visited Artek, while Soviet children came to Waterbury, Connecticut. The trades went well, and everyone seemed satisfied.

In retrospect, my 1989 journey to the Soviet Union was a profound and eye-opening experience. It revealed a society caught between its storied past and an uncertain future, struggling with the realities of economic decline, social hardship, and fading ideological fervor. The contrast between the grandeur of the Moscow Metro and the dilapidation of everyday life, the resilience of the people amidst scarcity, and the stark divide between the privileged few and the struggling majority all painted a vivid picture of a nation on the brink of transformation. My experiences there not only informed my professional decisions but also left me with lasting impressions of a world that would soon change forever.

How Northshore residents helped crack the Enigma codes

Submitted to the Gloucester Daily Times, but not yet published.

During World War II, the Nazi forces employed an ingenious machine that was used to encode all their military command and control messages. The machine, called Enigma, had 159 quintillion (that is 159 followed by 18 zeros) possible settings. And the code was changed every twenty-four hours. The Nazis were therefore confident that their encryptions were unbreakable. Most cryptologists agreed. *They were all wrong.*

Early in the war, a group of British mathematicians, based at Bletchley Park, approximately 50 miles outside London, under tremendous pressure to solve the problem, successfully broke the codes. The 2014 movie "The Imitation Game" presented a highly fictionalized account of the events.

In the film, the University of Cambridge mathematician Alan Turing is portrayed as the primary genius behind the project. In real life, he played a significant role, *but it was a team effort, and other mathematical geniuses also made substantial contributions.*

One of those geniuses came to America after the war and lived on Water Street in Newburyport for many years. His name was Gordon Welchman. Like Turing, he was also a mathematician at the University of Cambridge and, in 1939, was among the first four recruits to work at Bletchley Park.

The cracking of the Enigma codes began in Poland several years earlier. The Cipher Bureau of the Polish Government, under Dr. Marian Rejewski, also a mathematician, developed an early form of a computer, which they called a Bomba, named after an ice cream dessert, to help break the Enigma codes. When the Nazis invaded their country, the Poles could not continue with their code breaking work, so they got their information out to England. It ended up at Bletchley Park.

Alan Turing began with the Polish design and developed his first computer, which he called a Bombe —a variation on the name of the Polish machine. This computing engine, completed in March 1940, shortened the time to break the daily code, but it was still not fast enough. It had to solve in much less than twenty-four hours because that is when the Nazis changed their Enigma settings.

It was Gordon Welchman, the later Newburyport resident, who came up with a technical solution to make the Bombe operate considerably faster. He added a Diagonal Board to the design of Turing's rudimentary computer, which dramatically sped up the machine's processing time. Now, the Bletchley team could find the key to the daily Enigma code settings in an hour or two and start using their own Enigma machines to read that day's Nazi messages.

Welchman, while at Bletchley, was also credited with another significant advance in cryptoanalysis. He called it "traffic analysis". In his approach, Welchman collected data-on-the-data (metadata) that was sent in the Nazi messages. He then used sophisticated, mathematical analyses to draw conclusions about the enemy's activities. It was an especially useful approach *and is still employed by our National Security Agency (NSA) as well as the British equivalent, the Government Communications Headquarters (GCHQ).*

Winston Churchill credited Welchman and his associates at Bletchley Park for having shortened World War II by at least two years. In the 1944 King's Birthday Honors list, *Welchman was awarded the Most Excellent Order of the British Empire (OBE).* Of course, his work was so secret at the time that it was not revealed why he had obtained the high honor.

Turing and Welchman were key figures at Bletchley Park, but the effort involved literally thousands of people, many of whom were women. Most of these women were members of the Women's Royal Naval Service (WRNS). *The British called them Wrens.*

While not shown in the movie, eventually, hundreds of the top-secret, Turing Bombe machines were built by the British

Tabulating Machine Company. Many were placed at Bletchley Park, while others were set in various locations around Bletchley to distribute the risk of losing the machines in the event of an air raid. All the Bombes were operated around the clock by teams of Wrens.

Each day, the code breakers at Bletchley would identify "cribs", or best guesses of encoded words, from that day's messages. For example, many messages ended with the words, "Heil Hitler." The codebreakers provided these cribs to the Wrens, who would program the Bombes using this information. One Wren would set the drums on the front of the Bombe, while the other would plug the wires in the rear of the machine to their appropriate settings. If the crib information were correct, the machine would eventually stop running, and the Wrens could read the Enigma settings for that day. So, the Wrens played a key role in the work at Bletchley Park.

Two of these Wrens eventually moved to the United States and lived on the North Shore. One was Diana Lucy of Newburyport, and the other was Mary Wittkower of Rockport.

Diana also lived on Water Street in Newburyport, not far from Gordon Welchman. Yet they did not know each other. They found their connection when they met at a dinner party given by Diana for her neighbors. She was told that Gordon worked at Bletchley Park during the war. And so did she. They were delighted to find each other so far away from home.

Mary Wittkower of Rockport told her story in a fascinating memoir, "**An Extraordinary Time: Memories of a Young Girl Growing Up in WWII**". She enlisted in the WRNS as soon as she turned seventeen in 1944. The young girl was sent for basic training to a facility in Glasgow, Scotland. While there, she was selected for a "special assignment". The task she discovered later was to become a Wren Bombe operator. She was stationed at Eastcote, a facility not far from Bletchley Park, where eighty Bombes were located.

Then, on May 7th, 1945, Germany surrendered to the Allies. The war was over. Seeking new challenges, Gordon Welchman left England for the United States. After a stint as a computer instructor at MIT, he joined the MITRE Corporation in Bedford, Massachusetts. MITRE was engaged in research for the National Security Agency. In his position, Welchman made some significant contributions to U.S. security during the Cold War and was recognized by his colleagues as a genius.

The mathematician retired from full-time employment in the early seventies and moved to Newburyport. He continued as a consultant to MITRE. But that situation came to an unhappy end by the mid-eighties.

In his retirement, Welchman noticed that others from Bletchley Park had written books about their experiences there. Consequently, he decided to write his own detailed account of his accomplishments, "The Hut Six Story: Breaking the Enigma Codes".

It was a big hit as a book, but both the National Security Agency and its British equivalent, the Government Communications Headquarters, were extremely upset by what Welchman had revealed. *His work in traffic analysis was still top secret, and he had discussed it in the book.* They never charged him with a crime, but they made him feel miserable in his final years. He died in Newburyport, October 1985, at the age of 79.

Gordon Welchman never received the same level of public recognition as Alan Turing. Still, scholars who have studied the work done at Bletchley Park certainly recognize his significant contributions to breaking the Enigma codes and shortening World War II. And now, so do you.

Update of the article

More on Gordon Welchman's biography - Gordon Welchman was born on June 15, 1906, in Bristol, England. A mathematician by training, Welchman studied at Trinity

College, Cambridge, where he developed *a reputation for his sharp analytical mind.*

In 1939, as the world stood on the brink of war, the 33-year-old Welchman was one of the first four mathematicians to be recruited to join a top-secret group of codebreakers at Bletchley Park, the British government's codebreaking center.

Welchman was assigned to lead Hut 6, one of the key sections of Bletchley Park tasked with breaking German Army and Air Force Enigma messages. He recruited and managed a team of talented cryptanalysts, mathematicians, and linguists, ensuring effective collaboration. Welchman quickly *distinguished himself not only as a brilliant cryptanalyst but also as an exceptional organizer.*

He recognized that the complex process of breaking Enigma codes required a systematic approach. Under his leadership, Hut 6 developed an efficient workflow, turning intercepted enemy messages into actionable intelligence at an unprecedented speed.

But *Welchman's most significant contribution came in the form of a groundbreaking innovation known as the "Diagonal Board."* This enhancement to Turing's Bombe machine significantly improved its capabilities. The Diagonal Board allowed the Bombe to test multiple logical relationships simultaneously, vastly increasing its speed and effectiveness.

After the war, Welchman continued his work in computing and intelligence, but his contributions remained a secret until he published his memoir, **The Hut Six Story**, in 1982. This publication, however, sparked a security controversy, and *the British government accused Welchman of revealing classified information, tarnishing his legacy.*

Gordon Welchman passed away on October 8, 1985, but his impact on the war effort and the field of cryptography is undeniable. As history continues to recognize the heroes of Bletchley Park, Welchman's name now stands alongside Turing's as one of the great minds who changed the course of history.

Conclusion to Part VI

In this concluding chapter of "Science, War, and Invention: Lives That Changed the World," we reflect on the extraordinary stories we have explored, tales of visionaries whose genius and determination transformed the world. From the daring exploits of Yuri Tuvim, whose journey as a Soviet dissident to an American patriot exemplifies the courage of his beliefs, to the brilliant minds who played pivotal roles in decoding the Enigma machine during World War II, the narrative of human ingenuity has been vividly portrayed.

We began with Dr. Isaac Adams, delving into the controversy over who truly invented the electric light bulb. This story illuminates the complexities of scientific credit and the relentless pursuit of knowledge. The remarkable achievements of Alexander Graham Bell on the North Shore remind us of how groundbreaking communication technology emerged from seemingly humble beginnings.

J. Robert Oppenheimer's dual legacy as the father of the atomic bomb and a moral philosopher who questioned the consequences of his own creation highlights the ethical dimensions of scientific progress. General James Gavin parallels his story. "The Jumping General," whose strategic brilliance shaped World War II, but whose reflections on leadership and responsibility resonate far beyond the battlefield.

The Hammonds of Gloucester further demonstrate that innovation is not just the realm of great scientists and academics but can also be the legacy of industrious and visionary families. Their contributions to engineering and technology have left a lasting mark.

Together, these stories illustrate a fundamental truth: science and invention are never purely technical pursuits. Whether in times of war or peace, the impact of these brilliant individuals continues to shape our world.

As we close this part of the book, we are reminded that the story of science is never truly complete. It is a journey, a continuous narrative shaped by the ambitions, fears, and dreams of those who dare to ask, "What if?"

A Final Word

As we reach the final page of *The Curious Mind*, we find ourselves standing at the crossroads of wonder and knowledge. Together, we have journeyed through the transformative power of artificial intelligence, the hidden depths of animal cognition, the cosmic possibilities of non-human intelligence, the volatile forces shaping our planet, the profound frontiers of medicine and metaphysics, and the stories of remarkable individuals whose ideas changed the course of history.

Rather than just a catalog of facts, each chapter has been an invitation to think more deeply about the world around us. We have seen how technology can enhance or threaten humanity, how animals can teach us about intelligence in surprising ways, and how the search for extraterrestrial life expands our sense of what is possible. We have explored how nature's power is both creative and destructive, how modern medicine stretches the limits of life, and how human ingenuity has repeatedly reshaped our destiny.

But if there is one lesson that **The Curious Mind** has consistently emphasized, it is this: curiosity is the engine of progress. It is the spark that has driven every scientific breakthrough, every philosophical debate, and every act of discovery. It is a uniquely human trait, one that has propelled us from the darkness of ignorance to the light of understanding.

Yet our journey is far from over. The questions we have explored are but stepping stones, leading to new mysteries waiting to be uncovered. The frontiers of knowledge are ever-expanding, and with each answer, new questions arise.

So, as you close this book, I encourage you to keep your curious mind alive. Ask questions, seek answers, and never be afraid to explore the unknown. After all, the most extraordinary adventures in life are those driven by a restless curiosity.

Anthony J. Marolda.

The End

Index

www.ingramcontent.com/pod-product-compliance
Ingram Content Group UK Ltd.
Pitfield, Milton Keynes, MK11 3LW, UK
UKHW020128290926
15471UKWH00003B/344